Issues in Space

The 'Issues in Space' series brings together a broad range of authors and disciplines to promote a diversity of views, perspectives, and critical approaches in the space sector. This series is designed to influence the development of the domestic and global space industry while it is in its formative stage, rather than trying to redress imbalances once they are entrenched in the industry.

This series aims to particularly promote the work of authors from traditionally marginalised groups, including women, early career researchers, and others; voices that are important in discussions of the development of the space industry. The scope of this series covers a broad range of key areas and disciplines that will be explored. Fundamental areas such as: outer space, space law, space policy, space regulation, space archaeology, space ethics, space situational awareness, space traffic management, military activities in space, commercial space, dispute resolution for space activities, space courts, space security, sustained human presence in space, space resource utilisation, space governance, space settlement, armed conflict in space, space resource utilisation, space manufacturing, space tourism, private actors in space, and space mining.

Issues in Space is an interdisciplinary series in that it is devoted to space scholarship but with the added unique focus on the promotion of diversity of authors, disciplines and opinions.

More information about this series at http://www.springer.com/series/16523

Melissa de Zwart • Stacey Henderson
Editors

Commercial and Military Uses of Outer Space

Springer

Editors
Melissa de Zwart
Adelaide Law School
The University of Adelaide
Adelaide, SA, Australia

Stacey Henderson
Adelaide Law School
The University of Adelaide
Adelaide, SA, Australia

ISSN 2662-902X ISSN 2662-9038 (electronic)
Issues in Space
ISBN 978-981-15-8923-2 ISBN 978-981-15-8924-9 (eBook)
https://doi.org/10.1007/978-981-15-8924-9

This Springer imprint is published by the registered company Springer Nature Singapore Pte Ltd.
The registered company address is: 152 Beach Road, #21-01/04 Gateway East, Singapore 189721, Singapore

Preface

This book is the result of a workshop held at the University of Adelaide in 2019, which sought to promote the scholarship of women who are working and researching in space. This book brings together chapters on law, science, archaeology, defence, policy, and more, all with a focus on space. The authors are drawn from Australia and overseas, from academia, government, industry, civil society, and the military. We are grateful for financial support from the Space Industry Association of Australia and the University of Adelaide which enabled the *Women in Space* workshop.

We would like to thank Claudia Floreani for her excellent research assistance.

Adelaide, SA, Australia — Melissa de Zwart

Adelaide, SA, Australia — Stacey Henderson

Contents

Contributors

Melissa de Zwart is Professor and Dean of Law, Adelaide Law School, University of Adelaide. With a background in legal practice and academic scholarship in the legal issues emerging from the development, use and commercialisation of cutting edge technology, she has published extensively on legal issues arising from outer space, surveillance, Big Data, social networking, virtual communities, the internet and autonomous vehicles. Dr de Zwart is the Deputy Chair of the Space Industry Association of Australia, a frequent media commentator on matters affecting civil, commercial and military uses of outer space and a contributor to many space projects, including the *Woomera Manual on the International Law of Military Space Activities*, the Manfred Lachs Space Law Moot and the Space Security Index. She is a Fellow of the Australian Academy of Law.

Alice Gorman is Associate Professor, Flinders University. Dr Gorman is an internationally recognised leader in the emerging field of space archaeology. Her research on space exploration has been featured in National Geographic, the Monocle, and Archaeology magazine. She is a faculty member of the International Space University's Southern Hemisphere Space Program in Adelaide. Dr Gorman is a Director on the Board of the Space Industry Association of Australia, a member of the Australian Institute of Aboriginal and Torres Strait Islander Studies, and a Councillor of the Anthropological Society of South Australia.

Laura Grego is a senior scientist in the Global Security Program at the Union of Concerned Scientists. She focuses her analysis and advocacy on missile defense, outer space security, and nuclear weapons. She has authored or co-authored numerous papers on a range of topics, including cosmology, space security, and missile defense, and has testified before Congress and addressed the UN General Assembly and the UN Conference on Disarmament on space security issues. Before joining UCS, Dr Grego was a postdoctoral researcher at the Harvard-Smithsonian Center for Astrophysics. She earned a PhD in experimental physics at the California

Institute of Technology, and a BS in physics and astronomy at the University of Michigan.

Kathryn Robison Hasani is a PhD Candidate at the University of Alabama in Political Science in the fields of American Politics and International Relations, with an outside minor in Communication. She also holds degrees in Anthropology, Near Eastern Studies, and American Studies from the University of Arizona and Youngstown State University. She has traveled the world for research and language studies, and is a poet and podcaster. Her research interests are in the fields of space policy and communication (both political and science) with a particular interest in the role communication plays in the formation and dissemination of national space policies. She is also a science communications consultant for STEM professionals and students and currently teaches Political Science at the University of Alabama and Wake Technical Community College.

Stacey Henderson is a Lecturer at Adelaide Law School, University of Adelaide. Her research focuses on the protective capacity of international law, the responsibility of States and governance of outer space and space technology. She is the Rapporteur and project manager of the *Woomera Manual on the International Law of Military Space Activities*, and has published widely on issues of space law. Dr Henderson is a member of the International Institute of Space Law, the International Law Association, and the Space Industry Association of Australia, and a Council member of the Australian and New Zealand Society of International Law.

Elsbeth J. Magilton is the Executive Director for Technology, Security, and Space Law Initiatives for the University of Nebraska College of Law where she oversees multiple programs and developing initiatives, including the nationally recognized Space, Cyber, and Telecommunications Law Program. She is the law college's principal researcher for projects serving the U.S. Strategic Command's University Affiliated Research Center – the only such center in the Depart of Defense to include a law school in their research partnerships. She has been quoted in The Atlantic, The Verge, Space.com, a host of regional publications, and was a booked speaker at SXSW in March 2020. Ms Magilton guest lectures and speaks internationally on military-academic partnerships, space law and security, women in technology, and technology related issues. In September 2018 Ms Magilton and Matt Schaefer received a major NASA Space Law pilot-program grant to create a nationwide network of students, faculty, and practitioners interested in space law and policy. She is admitted to practice law in the state of Nebraska, U.S., and is member of the Nebraska Bar Association and the American Bar Association, including the ABA Air and Space Section. Ms Magilton is a member of the American Society of International Law and serves at the Co-Chair for the Space Law Interest Group. She is a member of the Legal Expert Pool for the European Centre of Excellence for Countering Hybrid Threats, and sits on the editorial board for the U.S. Air Force Academy Eisenhower Center for Space and Defense Studies Journal.

Samantha Le May is a PhD Candidate and Fulbright Scholar based at RMIT University's SPACE Research Centre. Motivated by the need for space as a resource to be used sustainably, her research is focused on bridging the scientific and policy aspects of Space Situational Awareness and Space Traffic Management. She is an active member on the IAA Permanent Committee on Space Debris, and the IAF Committee on Space Security.

Sumen Rai spent a decade as a teacher and science communicator in both classrooms and science centres. She currently works in the South Australian public sector, facilitating research and development activity through connecting researchers, industry and government. In this role, Sumen established and managed the South Australian events for the NASA Space Apps Challenge, as well as being involved in other open government hackathons such as GovHack. Sumen is a member of the Board of SA Makers, a not-for-profit organisation which runs Makerspace Adelaide, and aims to engage, grow and promote a diverse community of makers in South Australia.

Rajeswari Pillai Rajagopalan is a Distinguished Fellow and heads the Nuclear and Space Policy Initiative at the Observer Research Foundation. She is also a Non-Resident Indo-Pacific Fellow at the Perth USAsia Centre from April to December 2020. She was also the Technical Advisor to a new United Nations Group of Governmental Experts (GGE) on Prevention of Arms Race in Outer Space (PAROS) (July 2018–July 2019). As a senior Asia defence writer for The Diplomat, she writes a weekly column on Asian strategic issues. Dr Rajagopalan joined ORF after a 5-year stint at the National Security Council Secretariat (2003–2007), Government of India, where she was an Assistant Director. Prior to joining the NSCS, she was Research Officer at the Institute of Defence Studies and Analyses, New Delhi. She was also a Visiting Professor at the Graduate Institute of International Politics, National Chung Hsing University, Taiwan in 2012. She has authored or edited nine books including Global Nuclear Security: Moving Beyond the NSS (2018), Space Policy 2.0 (2017), Nuclear Security in India (2015), Clashing Titans: Military Strategy and Insecurity among Asian Great Powers (2012), The Dragon's Fire: Chinese Military Strategy and Its Implications for Asia (2009). She has published research essays in edited volumes, and in peer reviewed journals such as India Review, Strategic Studies Quarterly, Air and Space Power Journal, International Journal of Nuclear Law and Strategic Analysis. She has also contributed essays to newspapers such as The Washington Post, The Wall Street Journal, Times of India, and The Economic Times. She has been invited to speak at international fora including the United Nations Disarmament Forum (New York), the UN Committee on the Peaceful Uses of Outer Space (COPUOS) (Vienna), Conference on Disarmament (Geneva), ASEAN Regional Forum (ARF) and the European Union. Dr. Rajagopalan tweets @raji143 and maintains a personal blog: http://securitystrategyrajagopalan.blogspot.com/.

Cassandra Steer is a Lecturer at the Australian National University College of Law, specialising in space law and international law, and a Mission Specialist with the ANU Institute for Space. She has over 14 years' experience in academia in Australia, Canada, The Netherlands and the United States, teaching space law, criminal law, international criminal law and the law of armed conflict. Her former appointments include Acting Executive Director of the Centre of Ethics and the Rule of Law at the University of Pennsylvania law School, Executive Director of the McGill Centre Research in Air and Space Law, and Executive Director of Women in International Security-Canada. Dr Steer has also consulted to the Canadian and U.S. departments of defence on issues of space law and space security.

Susan Trepczynski is an attorney in the United States Air Force. She received a LL.M. in space law from McGill University and has an extensive background in space law and operations. She has held several space-related positions, to include serving as a space and cyber operations law instructor. Susan has lectured in various academic environments, published on a variety of topics, and served as an external reviewer for several publications. She is currently one of the core experts involved in drafting the *Woomera Manual on the International Law of Military Space Activities.*

Chapter 1
Introduction

Melissa de Zwart and Stacey Henderson

Abstract This chapter introduces the diverse range of disciplines and scholarship brought together in this collection. It is hoped that the spectrum of viewpoints presented in this book will challenge some existing views and norms of space as distant, harsh and barren, and reflect instead the reality of this vast domain as a vibrant and vital subject for study, exploration and inspiration.

The space industry is growing exponentially, with high levels of government and private funding combining with advances in technology to generate new opportunities in space. Yet there is deep uncertainty with respect to many aspects of space activities, including what law (if any) applies in space and whether objects are space junk or valuable items of cultural heritage.

Space scholarship is frequently discipline specific or siloed into either commercial or military uses of space. This book takes a different approach – that there are just uses of space and that consideration of a diverse range of disciplines and scholarship is essential to the long-term sustainability and success of all space activities for the benefit of all humankind.

Since the establishment of the Australian Space Agency in 2018, there has been a dramatic increase in space sector activity in Australia. As Sumen Rai observes this growth in activity has been accompanied by a desire to increase public engagement with the Australian space sector and attract a diverse workforce for the sector. Rai argues that for these efforts to succeed, the dominant discourses must be critically examined for 'hidden' exclusionary messages and practices. Rai's chapter challenges policymakers, developers of education programs and space sector practitioners to think critically about their messages and practices, especially those around race, class and gender, as a first step towards implementing change.

Kathryn Robison Hasani addresses the question of what the future of space exploration and exploitation will look like as technology continues to outpace

M. deZwart · S. Henderson (✉)
Adelaide Law School, The University of Adelaide, Adelaide, SA, Australia
e-mail: melissa.dezwart@adelaide.edu.au; stacey.henderson@adelaide.edu.au

M. de Zwart, S. Henderson (eds.), *Commercial and Military Uses of Outer Space*, Issues in Space, https://doi.org/10.1007/978-981-15-8924-9_1

regulation and governance, and more actors gain space capabilities. In her chapter, Robison Hasani explores whether the future will resemble the utopian ideals of Star Trek, or whether it will devolve into a Star Wars-esque situation of colonialism and profiteering.

Discussing business case research to show the economic advantages of gender-diverse leadership, Elsbeth J. Magilton examines data from the United States, alongside individual personal interviews, to paint a picture of gender diversity in the national security and space sectors. She discusses women-focused professional organizations and female student interest in national and space security, as well as United States' space security policy and the roles women play in that sector's development. Finally, Magilton outlines efforts at the University of Nebraska to support gender diversity in national space security and how educators can modify their pedagogical approaches to promote gender diversity.

Susan Trepczynski suggests that understanding space law, and the interplay between international legal obligations and domestic law, is key to understanding the fundamental bases on which space activities are prioritized, integrated, and utilized in a given State. A State's domestic legal regime may have a significant impact on the development of space activities and emerging space-related technologies, which can be enabling, constraining or neutral. Trepczynski argues that for those actors who are developing emerging technologies, seeking to engage in innovative space activities, or advising those who do, it is critical to understand how the applicable legal regime will impact the activity or technology.

Through the Artemis program, NASA has invited participation of space agencies from across the world in crewed missions to the Moon and to Mars. However, as Melissa de Zwart notes, the opportunity to participate in the Artemis program is subject to compliance by partner agencies with the Artemis Accords, a series of principles which both restate and extend certain laws applicable to the exploration and use of outer space. In her chapter, de Zwart considers the effect of the Artemis Accords, addressing in particular the role of commercial space operators in a domain previously considered to be accessible only by States. She concludes that the Artemis Accords introduce some radical new concepts to space law at a critical moment in the evolution of the space industry.

Laura Grego evaluates arms control as a way to provide a future in space that is more secure and more sustainable and can be used to help create a more safe and secure future on Earth. In her chapter, Grego argues that a dangerous space arms race dynamic has been unfolding for decades. This has begun accelerating in recent years, with the ability to hold satellites at risk becoming more sophisticated and proliferating, both on the ground and in orbit. This risks the space strategic environment becoming more threatening and unpredictable and the space physical environment for all users becoming degraded because of fast-moving space debris. Grego argues that space arms control is urgently needed and can be achieved, given sufficient political will.

Stacey Henderson notes that there are very few rules for the resolution of disputes set out in international space law. Existing international space law contains only a fragmented, patchy and out-dated system for the resolution of disputes relating to space activities. Henderson outlines and evaluates the existing international dispute resolution mechanisms available to resolve space disputes and considers whether there is a need to establish a new *sui generis* international dispute settlement body to adjudicate disputes relating to space activities.

As earth-orbital environments become increasingly populated, Space Situational Awareness and Space Traffic Management capabilities will be key to coordinating international space activities. Samantha Le May argues that to preserve the long-term sustainability of the outer space environment, new approaches that can leverage diverse data from multiple information sources to generate new insights are needed to inform decision making in the space sector. In her chapter, Le May describes the current population of space objects (including satellites and space debris), details existing approaches to Space Traffic Management, identifies emerging challenges for the space sector, and demonstrates the value of graph databases as an analytic tool to inform decision making in the space domain.

Looking at space situational awareness in a different way, Alice Gorman argues that numerous defunct satellites in Earth orbit have cultural significance for populations of Earth for their historic, aesthetic, scientific and social values, which are not easily accommodated within the space situational awareness model. Gorman argues that effective heritage management is better pursued using an environmental management framework which incorporates the intrinsic values of the space environment as a new entity created by the synergistic effects of natural and cultural activities. Gorman elaborates on mechanisms and concepts such as heritage lists, themes and offsets, necessary for ensuring that some culturally significant spacecraft survive into the future.

Rajeswari Pillai Rajagopalan argues that the biggest challenge to the safety, security and sustainability of space continues to be the lack of consensus among major space powers which has impeded the development of an effective space security regime. Rajagopalan calls for a fresh approach to space security governance, with a pro-active role for middle space powers and solutions that are threat-specific, rather than overarching agreements.

Cassandra Steer argues that greater diversity is needed in the space sector, and will only be achieved when women feel they are truly part of the structures and institutions that govern space. Steer notes that while we cannot change the gender-specific language in the Outer Space Treaty, we can change the way language is used today within space governance power structures and how it is used in future legal instruments. Steer presents a feminist legal analysis of space law, revealing the power dynamics in the processes of law-making, interpretation and application which continue to exclude women and to maintain a status quo which includes geopolitical dominance by a small group of elite countries.

This book covers some of the diverse issues facing contemporary activities in space and those attempting to understand, use and regulate the space domain. By bringing together a group of authors who are all experts in their respective fields, this collection of works written by all female authors provides a significant contribution to scholarship on space and promotes diversity – in its broadest sense; diversity in terms of gender, discipline, career stage, and so on – both within the space sector and in scholarship relating to space activities. It is hoped that this spectrum of viewpoints will challenge some existing views and norms of space as distant, harsh and barren, and reflect instead the reality of this vast domain as a vibrant and vital subject for study, exploration and inspiration.

Chapter 2
Whose Space? A Critical Approach to Increasing Public Engagement with the Australian Space Sector

Sumen Rai

Abstract Space sector activity in Australia has been increasing rapidly since the Australian Space Agency was established in 2018. This growth has been accompanied by a desire to increase public engagement with the Australian space sector, enacted through formal, non-formal and informal education, and community outreach. This chapter argues that attempts to increase public engagement, including attracting a diverse workforce for the sector, will not succeed unless the dominant discourses which are being reproduced through public engagement activities (including those around race, class, and gender) are examined critically for 'hidden' exclusionary messages and practices. The aim of this chapter is to provoke policymakers, developers of education programs and space sector practitioners to think critically about their messages and practices, as a first step towards implementing changes.

As I work on a draft of this chapter, I am keeping an eye on two screens. On one, the first-ever commercial crewed launch is about to take two astronauts to the International Space Station. NASA Administrator Jim Bridenstine tweets about his pride at launching "American astronauts on American rockets from American soil". The other screen shows images of huge Black Lives Matter protests in cities across the US, with protestors enveloped in a haze of tear gas and nursing injuries from 'rubber bullets' shot directly into crowds. The juxtaposition is jarring: a space industry celebrating a billionaire entrepreneur, technological development, and national pride, while their fellow citizens protest for their basic rights to live free from the impacts of racism.

The views and opinions expressed in this article are those of the author personally and not her employer.

S. Rai (✉)
SA Makers & Makerspace Adelaide, Adelaide, Australia
e-mail: sumen@makerspaceadelaide.org

M. de Zwart, S. Henderson (eds.), *Commercial and Military Uses of Outer Space*, Issues in Space, https://doi.org/10.1007/978-981-15-8924-9_2

Responses to both the launch and the protests flood my social media feeds. People from around the world, including Australia, are sharing their experiences of racism and violence in solidarity with the protestors. I notice posts from people in the Australian space sector congratulating SpaceX and NASA on the launch . . . and I also notice, as the number of these posts grows, a disquieting silence about the protests and the discussions sparked by the protests.

2.1 Introduction

Space sector activity in Australia has been increasing rapidly since the Australian Space Agency was established in 2018, with the creation of new space businesses, increased research and development, international engagement, and growth in the number of space-related educational and outreach initiatives. This activity includes efforts to engage the Australian public through a variety of methods: 'formal education' including schools, universities and vocational education and training; 'non-formal education' including museums, science centres and initiatives such as space camps; 'informal education' including books, documentaries, and social media platforms such as Twitter; and 'community outreach' including public lectures, events and community consultations on policy and legislation.

The purpose of increasing public engagement is often framed around the need to build a future workforce for the sector. The Australian Space Agency sums it up in the *Advancing Space – Australian Civil Space Strategy 2019–2028:*

> By inspiring the next generation through the prism of space, Australia will increase its STEM educated population and create a workforce pipeline to sustain and grow a national space industry. This includes exploring new ways to engage with young people, working with successful entities locally, and drawing on global best practice (Australian Space Agency 2019).

There is also recognition of the need for the sector to become more inclusive and diverse: " . . . to foster innovation and creativity, we need a diverse workforce. This is why the Agency's values include being inclusive" (Department of Industry, Science, Energy and Resources 2019).

The statement above is an example of how attempts to be inclusive may have the opposite effect. By linking the drive for diversity and inclusion to economic outcomes (ie. innovation and workforce development) rather than, for example, to the social justice outcome of opening up opportunities for people who have historically been excluded from participating by discriminatory power structures, it sends the message that diversity will be valued insofar as it contributes to the existing power structures, leaving those power structures fundamentally intact.

This chapter argues that messages such as these, and the activities that are built around them, can end up working against the space sector's aims to increase public engagement. Without actively challenging these dominant discourses and practices, implementing an increasing number of engagement activities will not be effective in increasing public engagement with the space sector.

The issues outlined in this chapter are faced by other industry sectors too, but the Australian space sector is in the unique position of being in the early stages of development. Those who have influence over the development of the sector have an opportunity (and responsibility) to avoid replicating the exclusionary and discriminatory systems entrenched in more established sectors by 'building in' inclusive and equitable discourses and practices from the beginning.

The aim of this chapter is to provoke policymakers, developers of education programs and space sector practitioners to think critically about the messages they promulgate and the activities they implement. It does not provide easy solutions but hopes that the questions it raises will be the first step in the development of future policies, practices and initiatives where impactful changes can be implemented.

2.2 What Are the Dominant Discourses in the Australian Space Sector?

The dominant discourses in the Australian space sector serve to create, replicate and reinforce a set of ideological positions which can be described as 'neoliberalism', which tends to favour free markets, reduced government intervention, privatisation, and deregulation. Neoliberal hegemony is powerful because it allows for the inclusion of 'progressive' views such as increasing diversity and inclusion, as long as those views can be made to fit with the overarching imperatives, "... silencing complex inequalities and power relations" (Gyamera and Burke 2018).

Hegemony creates 'in-groups' and 'out-groups' based on the level to which people subscribe to, or serve, the dominant discourses: "In the neoliberal political economy of belonging, inclusion and exclusion are increasingly becoming a function of an individual's, or a group's, capacity to contribute to the country's financial viability, economic competitiveness, international reputation ..." (Mavelli 2018). However, a feature of hegemony is that it "... holds sway over a set of subjects by winning their consent or securing their compliance" (Howarth 2010), and this requirement to win consent and compliance implies that consent and compliance can be withdrawn. This possibility for resistance is powerful, but real change can only be affected once the boundaries between those included and excluded is redrawn, which will require fundamental change in the discourses and practices of the sector. Those who are currently in the 'in-group' must be motivated and capable of identifying who is being excluded by current dominant discourses and social practices, and then work on dismantling and replacing those discourses and practices.

2.2.1 The Space Sector Values Diversity and Inclusion

As a technology-intensive sector, much of the diversity and inclusion discourse in the space sector is about increasing diversity in STEM disciplines. For example, the Australian Space Agency aims to "Implement STEM initiatives and partnerships with industry to develop the future diverse and inclusive workforce, including Indigenous engagement and internships" (Australian Space Agency 2019).

The requirement for a STEM workforce, and how to convince more students to undertake subjects in STEM disciplines, are issues which have been concerning governments and a range of technology-intensive industries. In response, STEM policies and initiatives aimed at increasing participation in STEM disciplines have proliferated in Australia in the last decade, with a 2016 STEM Programme Index from the Office of the Australian Chief Scientist listing 250 initiatives (Commonwealth of Australia 2016), and a mapping exercise by the Australian Academy of Science (2018) finding 331 initiatives in Australia with the specific goal of increasing participation of girls and women in STEM. However, declining STEM participation rates in Australian schools and universities (Timms et al. 2018) raise the pressing question of why the increase in initiatives does not seem to be resulting in an increase in the uptake of STEM subjects.

The underlying factors for low participation rates in STEM disciplines are numerous and sometimes confounding. Some of these factors are societal, structural issues which schools and science initiatives cannot address on their own (Evans 2005). For example, there is evidence that low achievement in literacy, numeracy and general scientific knowledge in students who are starting school are predictors of low performance and engagement with STEM subjects in later years, and that these are linked with low socioeconomic circumstances and lack of access to well-funded schools and extracurricular enrichment activities (Morgan et al. 2016; Timms et al. 2018; Evans 2005).

Another structural issue which has particular resonance for the Australian space sector is the legacy of colonialism. Australia's first foray into the space industry began at the Woomera Test Range with a joint project between Australia and the United Kingdom to develop a long-range weapons testing facility. The site, located on the land of the Maralinga Tjarutja, Anangu Pitjantjatjara Yankunytjatjara, Antakirinja Matu-Yankunytjatjara, Kokatha, Arabana and Gawler Ranges people, was chosen because of its remoteness and quiet electromagnetic environment (Commonwealth of Australia 2018). It is important to remember that the Indigenous communities of the area were dispossessed of their lands and prevented from accessing significant sites within the WPA. While the Joint Anglo-Australian Project was being conducted at Woomera from 1947 to the early 1970s, there was still a program underway of assimilation of Indigenous people, and communities were dealing with the health, social and emotional implications of these policies, including the forced removal of Indigenous children from their families. This was also a period of increasing activism for the civil rights of Aboriginal people, but progress was slow. It took until 1967, the year Australia launched its first satellite, the Weapons Research Establishment Satellite (WRESAT), for a referendum to be held to amend the Constitution so Aboriginal people could be counted as part of the population in the census (Dow and Gardiner-Garden 2011).

The Woomera Prohibited Area continues to be a contested site, with a co-existence framework involving the Australian Department of Defence, mining companies, pastoralists and Aboriginal groups. While Aboriginal groups currently have either freehold titles or native title over areas within the Woomera Prohibited Area, the story of their initial dispossession and resistance to development, set within the systemic atrocities committed against Aboriginal people and the enduring legacy of colonial violence, is rarely mentioned in detail when official Australian space history is told.

The Australian space sector is taking initial steps to increase Indigenous engagement and participation. For example, the Australian Space Agency's brand was created in consultation with Indigenous communities and experts in Indigenous astronomy, and its logo incorporates constellations which embody stories and meanings for a number of communities across Australia. The Australian Space Agency's strategy also mentions Indigenous engagement and internships specifically, although it is unclear at this early stage how this will be implemented. From the broader space sector, there are a number of initiatives across industry and education including: Australia's first Indigenous-built and owned satellite ground station, on Aboriginal owned land in the Northern Territory (Indigenous Business Australia 2020); the development and operation of the Arnhem Space Centre, a civilian launch facility developed in partnership with Indigenous-owned Gumatj Corporation (Developing East Arnhem 2019); the establishment of the Australian Indigenous Education Program at the University of Technology Sydney (UTS), which funds Indigenous students to undertake educational programs at NASA's Jet Propulsion Laboratory (UTS 2019); and the continuing rise of Deadly Science, an Indigenous-owned science education and outreach business which is providing books, equipment such as telescopes, and mentoring to students in remote Indigenous communities (Deadly Science 2020).

These are positive steps towards increasing Indigenous engagement in the space sector. However, diversity and inclusion efforts must not stop at marketing and individual educational initiatives, but need to be sustained through the questioning of every assumption the space sector makes in discourse and practice.

There is perhaps no more obvious example of hegemonic discourses creating 'in-groups' and 'out-groups' in the space sector than the increasing privatisation of space activities Holland and Burns (2018), captured under the umbrella term 'NewSpace'. Billionaire entrepreneurs such as Elon Musk, Jeff Bezos and Richard Branson have entered the space market with vast resources to channel toward development of novel space launch and transport technology, exploration, tourism, and provision of services through deployment of megaconstellations. These entrepreneurs are often portrayed as risk-taking visionaries who will succeed at introducing transformative technologies and business models for the space sector globally. Privatisation of activity in space is viewed as a positive development within the hegemonic discourses of the space sector, but while the rhetoric is that space is becoming more accessible (or in the words of NewSpace proponents, 'democratised'), in reality, it still requires a high level of skills and resources and so will remain the province of those who have access to the appropriate education and capital (as will the rewards). Activities under the NewSpace paradigm are undertaken not for humankind but for 'capitalistkind' (Shammas and Holen 2019).

On a smaller scale, the Australian space sector is also focusing on the development of startups and scaleups, and some of the rhetoric around the billionaire entrepreneurs described above has been projected onto Australian founders. While these Australian founders and their companies provide a more accessible version of NewSpace, creating successful startups requires access to social capital such as educational attainment and a business network, and financial risk-taking which is not an option available to everyone. Instead of inspiring more people to become entrepreneurs, the space sector risks alienating those who prefer to participate in other ways.

2.2.2 Space Unites Humanity

The newfound ability for humans to create images of Earth from space in the twentieth century inspired poetic reflections on the unity and fragility of humankind, expressed perhaps most famously by Carl Sagan describing the 'Pale Blue Dot' image taken by Voyager 1, when it was about 6.4 billion kilometres away from Earth:

> Look again at that dot. That's here. That's home. That's us. On it everyone you love, everyone you know, everyone you ever heard of, every human being who ever was, lived out their lives . . . every saint and sinner in the history of our species lived there – on a mote of dust suspended in a sunbeam (Sagan 1994).

This perspective is similar to the 'overview effect' reported by some astronauts, and described by author Frank White in his 1987 book *The Overview Effect: Space Exploration and Human Evolution*:

> That experience of seeing the Earth from orbit, or from the Moon, and having a realization of the inherent unity and oneness of everything on the planet. It's a realization that we are all one in terms of our place in the universe and our destiny . . . It's a shift in consciousness, a shift in awareness, and identity, and a harbinger of many more evolutionary transformations (quoted in Bimm 2014).

At the time White's book was written, there had been a growing environmental movement for two decades, popularising conceptions of the Earth as a delicately balanced, interconnected and self-regulating closed system as represented by Buckminster Fuller's Spaceship Earth in the 1960s, the 'Blue Marble' image of Earth captured by the Apollo 17 astronauts, and James Lovelock's Gaia Hypothesis in the 1970s (Bimm 2014). Some recurring themes in current descriptions of the overview effect include a connection to the Earth, an understanding of the interconnectedness of everything on Earth, and strong emotional responses to seeing negative impacts such as deforestation and desertification (Voskia 2020). The danger of these 'high-level' perspectives on the Earth is that they erase the nuances of individual and community struggles against racism, dispossession, poverty, and other forms of injustice.

It is likely that astronauts' interpretations of the cognitive shifts they experience as they view Earth from space are informed by the cultural references familiar to them, but since the pool of astronauts is limited to a few individuals from very few countries, it is unclear if astronauts from diverse cultural contexts and backgrounds would experience the overview effect differently. There are lessons in this for the broader space sector if it is to become more inclusive, in that any assumptions that interest in, and experiences of, space are a shared human condition should be challenged.

The unity of humanity in the context of outer space is also reflected in international law, where outer space is considered a 'global commons'. The United Nations' Outer Space Treaty and Moon Agreement use the terms "common province of mankind" and "common heritage of mankind", but the interpretation and application of these concepts continues to be a source of much debate (Kerrest 2011). As an increasing number of nations and private sector organisations achieve access to space, the aims of the Outer Space Treaty and Moon Agreement are likely to be tested because of the reputational and financial rewards potentially on offer.

This raises important questions about who can decide which activities in space are acceptable on behalf of the rest of humanity. There are some examples in recent years where the unilateral activities of states or the private sector have caused consternation. For example, SpaceX's Starlink satellite constellation was already shown to be interfering with astronomers' ability to observe the sky when the company launched their first tranche of 60 satellites, as the reflective objects were creating streaks in astronomical images. Numerous companies are planning to launch megaconstellations in the future, and SpaceX alone has plans to launch around 12,000 Starlink satellites (Grush 2020; Resnick 2020).

The United States' Executive Order on Encouraging International Support for the Recovery and Use of Space Resources (2020) states that the US does not consider space to be a global commons, raising questions about international treaty obligations, territorial claims and commercial exploitation of resources in space (Storr 2020; Freeland and Handmer 2020; Stimers 2020); and the establishment of a new branch of the United States Armed Forces, the Space Force (Goel 2020) adds to the ever-present geopolitical tension about militarisation and weaponisation of space.

While the ability for the general public to influence these global developments is limited, one avenue for them to be heard is through engagement with their government representatives. Consulting with diverse groups in the Australian community, as well as increasing the diversity in senior government positions will lead to responses to geopolitical situations which are more representative of the diversity of views of the Australian population.

Another way that space supposedly unites humanity is through the shared earthly benefits space programs provide. NASA even has a website dedicated to highlighting technological spinoffs from the space program (NASA 2020). Despite this, criticism of the costs of space activity, particularly human spaceflight, continue. The most compelling of these criticisms is not about space activities *per se*, but about the priorities of governments when it comes to allocating public sector effort and public money. This was powerfully expressed by Gil Scott-Heron (1970) in 'Whitey on the Moon', in which he described the ongoing structural injustices of poor healthcare, poor education, inadequate housing, and food poverty faced by Black people in the US, while at the same time, their government was focused on the expensive and non-essential task of putting humans (specifically, white men) on the Moon.

The Apollo program sparked protests across the United States about the expense of human spaceflight and the priorities of the US government, with only around half of Americans supporting the program (Madrigal 2012). As one example of resistance to the expense of the space program, former Chief Historian of NASA, Roger Launius, describes an encounter between NASA Administrator Thomas Paine and Reverend Ralph Abernathy, who took over as head of the Southern Christian Leadership Conference after the assassination of Dr. Martin Luther King, Jr., and had led a protest to the Apollo 11 launch site at Cape Canaveral. Paine described his conversation with the activists:

> I stated that if we could solve the problems of poverty in the United States by not pushing the button to launch men to the moon tomorrow then we would not push that button . . . I said that although I could not promise early results, I would certainly do everything in my own personal power to help him in his fight for better conditions for all Americans, and that his request that science and engineering assist in this task was a sound one which, in the long run, would indeed help (quoted in Launius 2014).

It is a familiar argument from space advocates that activities in space will have benefits on Earth. Fifty years after Apollo 11, the description above reads as a powerful white man paternalistically dismissing the voices of Black people who wanted to raise valid criticism of the priorities of their government.

2.3 Replication of Hegemonic Discourses Through Public Engagement Activities

A common response to the challenge of increasing engagement and participation in a sector is to create new education and outreach initiatives. For example, in response to a predicted shortfall in STEM-qualified workers in Australia for a range of technology-intensive sectors, initiatives to attract and retain young people in STEM disciplines proliferated in the last decade. Despite the effort to create and deliver novel educational programs and outreach initiatives, however, STEM participation rates continue to be low.

What seems to be missing is a recognition that education and outreach replicate the hegemonic discourses of the society in which they exist unless they are actively challenged (Asen 2017; McGregor 2009), and that education and outreach practices can themselves create barriers to engagement and participation. ‘Cultural mismatch theory’ shows that “when institutions promote only mainstream norms, they inadvertently fuel inequality by creating barriers to the performance of underrepresented groups” (Stephens and Townsend 2015) and create emotional distress and psychological stress responses in individuals in these situations (Phillips et al. 2020; Sladek et al. 2020). The preceding sections have outlined some of the dominant discourses present the Australian space sector. Conducting greater numbers of activities while leaving these discourses unchanged will continue to create in-groups and out-groups, and limit the ability of the space sector to increase public engagement (Kliewer 2013).

While the phrase ‘public engagement’ is a convenient way to group education and outreach activities, it should also be challenged - the default concept of ‘the public’ in a neoliberal hegemony “upholds a market populated by atomistic individuals as a singular and universal sphere of activity” which “imposes a homogeneity on market actors, ascribing to them uniform motivations and goals” (Asen 2017). In order to achieve engagement by diverse groups with differing motivations and goals, it is useful to think in terms of multiple publics rather than a single entity called ‘the public’; to recognise that it is possible for individuals to participate in more than of those publics simultaneously; and to accept that these publics will bring different knowledges to any engagement. This final point is captured by the term ‘civic epistemologies’, which is a way of understanding public engagement which moves “away from *a priori* assumptions about what publics should know or understand” (Jasanoff 2007).

The public engagement ecosystem includes a wide range of activities, including formal, non-formal and informal education, and community outreach. There is a rich body of literature analysing the ways in which educational institutions and practices reproduce dominant discourses and structural social inequalities of class, gender, race (and others) across all levels of formal education (see Parkinson and Jones 2018; Naidoo 2004; Kapferer 1986 as examples). This occurs through both the skills and knowledge deemed as ‘worthy’ of inclusion in the formal curriculum (Ditchburn 2012), and what is described as the ‘hidden curriculum’, the often unexamined mix of mores, values, beliefs and prejudices which are present in classrooms and the social environment of a school. The hidden curriculum can reinforce or subvert the overt messages the school might be promulgating – in much the same way as the nuances of dominant discourses can have the opposite effect to their stated intent.

The key issue with non-formal and informal education practices is that they have long been predicated on the ‘deficit model’, that is, an assumption that a lack of societal support engagement with science and technology is caused by ignorance or scientific illiteracy (Simis et al. 2016; Trench 2008). The assumptions of the deficit model have led to great effort in the popularisation of science, and increase in science communication activities, often privileging novel ways of communication of scientific concepts and ignoring other factors which might explain a lack of engagement. This end result of this is an increase in the number of initiatives and activities without a corresponding increase in engagement, as is the case in Australia. Critical analyses of science centres and museums have examined the ways in which these institutions reproduce inequalities of gender (Rodéhn 2019), class, and race (Dawson 2014). Formal education has similar issues, but trained teachers and facilitators can play a role in guiding learners in how to interact and resist these dominant discourses. Dawson (2014) points out that in the absence of these guides in non-formal and informal contexts, "... learning hinges to a greater degree on individual, family or peer literacy and ability to grasp the 'right' thing to do when faced with an exhibit as well as the extent to which the exhibit is designed to be accessible. Knowing what to do hinges, in turn, on the previous experiences afforded by someone's social position and lifestyle".

The final avenue of public engagement is community outreach, including promotional activities, public lectures, events and consultations. Community outreach tends to share the 'deficit model' outlined for non-formal and informal education, with the bulk of activity for space agencies being the promotion of the value of the sector to garner public support, and to promote the sector to teachers and parents who will have an influence on the young people who will comprise the future workforce. The assumption when engaging with the wider community is that "... as long as people are properly informed about scientific facts, there should be no cross-cultural variation in their perceptions of science or their receptivity toward technology ... lack of understanding, in turn, becomes a deficit that states can correct through better dissemination of knowledge" (Jasanoff 2007).

The assumption that more information will lead to greater understanding is contradicted by evidence of an ongoing public engagement issue from the perspective of NASA: "For people in the space community, the space program means many things: jobs, money, progress, political capital, and prestige ... And for 50 years, people outside the space community have not been clear about the purpose of the space program. The rationales that NASA has offered over the years do not appear to be especially meaningful to the Agency's "external" audiences." (Billings 2010). The development of the Australian space sector is occurring under very different historical and political circumstances, but there are nevertheless lessons in this about not taking public understanding and support for granted, and ensuring engagement with the public goes beyond the sharing of information to something which is meaningful for them.

With growing expectations from citizens that government agencies will create opportunities for participatory democracy (Hendriks et al. 2019; Vromen 2003), there are increasingly activities which move away from the 'deficit model' and towards a two-way 'dialogue model' (as described in Trench 2008) such as community consultations on proposed legislation or strategy such as the Australian Space Agency's 'Moon to Mars Program Design' consultation. However, participating in these activities requires people to have the requisite skills, knowledge and time, so while they do not explicitly exclude participation from the 'general' public, in effect, they only involve one type of public – those who are already engaged and interested enough to participate.

Another avenue for public engagement which relies on the skills, knowledge and time of participants is the space-focused 'hackathon' or 'innovation challenge'. This competition format gives participants a limited timeframe to devise solutions to a set of problem statements (sourced from end-users such as government and industry) using space-derived data. Hackathons and innovation challenges have proliferated globally in the last decade. In Australia, even with its relatively small space sector, there are currently three annual space-focused hackathons and innovation challenges: the Deloitte Gravity Challenge, French Space Agency CNES' Act in Space and the NASA International Space Apps Challenge. The NASA International Space Apps Challenge is unique in this field in that it focuses more on 'public good' and less on commercial outcomes such as the creation of startup companies. Event-based education and innovation activities tend to be favourably received by the participants, and while there are questions about the sustainability of results from these types of activities, they nevertheless provide an effective avenue for popularisation of space (Juraschek et al. 2020; Rai and Griffiths 2016; Senghore et al. 2015).

Issues with the various types of public engagement outlined above will not be addressed quickly or easily. For example, disrupting exclusionary messages in the hidden curriculum in a school requires changing attitudes and beliefs of school administrators, teachers and students – which in turn requires changes in the broader hegemonic discourses which shape their attitudes and beliefs. The problem can seem intractable, but hegemony relies on consent and compliance and the possibility withdrawing consent and compliance, of examining the hegemonic discourses and actively working to change them.

2.4 Increasing Public Engagement with the Australian Space Sector

This chapter starts from the position that increasing public engagement with the Australian space sector is a valuable pursuit, not just to build a workforce in support of sector development, but also from a participatory democracy standpoint.

While it takes a critical look at dominant discourses and practices, the purpose of this chapter is not to suggest that we immediately discard all the current efforts being made by the Australian space sector to engage the public; rather, it hopes to outline some complexities and provoke those with influence over the development of the sector to think critically about their assumptions, and then to work on implementing genuinely inclusive discourses and practices. This is crucial if the sector wants to be successful in its aim to be diverse and inclusive, and to not replicate the issues faced by more established sectors.

Recognising that the nascent Australian space sector stands at a unique point in time when it diverse groups of people, some of whom might not have seen a place for themselves previously, if the sector can 'build in' inclusive discourses and practices from the beginning.

The subdued response of the Australian space sector to the Black Lives Matter protests reminds me that the Australian space sector is overwhelmingly white, and while it is becoming easier to speak about some forms of diversity and inclusion, such as increasing the participation of women in the sector, conversations about race are still difficult.

As the social media updates about both the launch and the protests continue, I start to see updates from Black scientists, engineers and other space professionals, some of whom I already follow and some whose posts have been retweeted and shared into my timeline in an effort to amplify their messages. The effect is powerful. The voices of those who have historically been marginalised now have a way of reaching an audience directly, without their messages being co-opted, sanitised and mediated to fit the hegemony.

We must listen and act.

References

Asen, R. (2017). Neoliberalism, the public sphere, and a public good. *Quarterly Journal of Speech, 103*(4), 329–349. https://doi.org/10.1080/00335630.2017.1360507.

Australian Academy of Science. (2018). *Mapping Australian STEM participation initiatives for girls and women.*

Australian Space Agency. (2019). *Advancing space Australian civil space strategy 2019–2028.*

Billings, L. (2010). Fifty years of NASA and the public: What NASA? What publics? In *NASA's first 50 years: Historical perspectives* (pp. 151–181).

Bimm, J. (2014). Rethinking the overview effect. *Quest: The History of Spaceflight Quarterly, 21*(1), 39–47.

Commonwealth of Australia. (2016). *STEM programme index 2016.*

Commonwealth of Australia. (2018). *Coexistence in the Woomera Prohibited Area 2018 Review.*

Dawson, E. (2014). Equity in informal science education: Developing an access and equity framework for science museums and science centres. *Studies in Science Education, 50*(2), 209–247. https://doi.org/10.1080/03057267.2014.957558.

Deadly Science. (2020). *Our mission.* https://deadlyscience.icu/. Accessed 15 July 2020.

Department of Industry, Science, Energy and Resources. (2019). *Balance for better in space.* https://www.industry.gov.au/news-media/australian-space-agency-news/balance-for-better-in-space. Accessed 20 June 2020.

Developing East Arnhem. (2019). *East Arnhem space capability profile 2019.* https://www.developingeastarnhem.com.au/assets/uploads/2019/11/East-Arnhem-Space-Capability-Profile-Final.pdf. Accessed 15 July 2020.

Ditchburn, G. M. (2012). The Australian curriculum: Finding the hidden narrative? *Critical Studies in Education, 3*(3), 1–4. https://doi.org/10.1080/17508487.2012.703137.

Dow, C., & Gardiner-Garden, J. (2011). *Overview of indigenous affairs: Part 1: 1901 to 1991.* Canberra: Deptartment of Parliamentary Services, Parliamentary Library.

Evans, G. W. (2005). Reframing the acheivement gap. *Phi Delta Kappan, 86*, 582–589.

Freeland, S., & Handmer, A. (2020). Giant leap for corporations? The Trump administration wants to mine resources in space, but is it legal? *The Conversation.* https://theconversation.com/giant-leap-for-corporations-the-trump-administration-wants-to-mine-resources-in-space-but-is-it-legal-136395. Accessed 30 June 2020.

Goel, B. (2020). The US space force and international law considerations. *The Space Review.* https://www.thespacereview.com/article/3879/1. Accessed 30 June 2020.

Grush, L. (2020). The true impact of SpaceX's Starlink constellation on astronomy is coming in focus. *The Verge.* https://www.theverge.com/2020/3/24/21190273/spacex-starlink-satellite-internet-constellation-astronomy-coating. Accessed 10 July 2020.

Gyamera, G. O., & Burke, P. J. (2018). Neoliberalism and curriculum in higher education: A post-colonial analyses. *Teaching in Higher Education, 23*(4), 450–467. https://doi.org/10.1080/13562517.2017.1414782

Hendriks, C. M., Regan, S., & Kay, A. (2019). Participatory adaptation in contemporary parliamentary committees in Australia. *Parliamentary Affairs, 72*(2), 267–289. https://doi.org/10.1093/pa/gsy005.

Holland, D., & Burns, J. O. (2018). The American space exploration narrative from the cold war through the Obama administration. *Space Policy, 46*, 9–17. https://doi.org/10.1016/j.spacepol.2018.03.007.

Howarth, D. (2010). Power, discourse, and policy: Articulating a hegemony approach to critical policy studies. *Critical Policy Studies, 3*(3–4), 309–335. https://doi.org/10.1080/19460171003619725.

Indigenous Business Australia. (2020). *Australia goes live with 1st Indigenous satellite ground station*. https://www.iba.gov.au/2020/07/australia-goes-live-with-1st-indigenous-satellite-ground-station/. Accessed 15 July 2020.

Jasanoff, S. (2007). Civic epistemology. In *Designs on Nature: Science and democracy in Europe and the United States* (pp. 247–271). Princeton, NJ, Princeton: Princeton University Press.

Juraschek, M., Büth, L., Martin, N., Pulst, S., Thiede, S., & Herrmann, S. (2020). Event-based education and innovation in Learning Factories–concept and evaluation from Hackathon to GameJam. *Procedia Manufacturing, 45*, 43–48.

Kapferer, J. L. (1986). Curricula and the reproduction of structured social inequalities. *Curriculum Inquiry, 16*(1), 5–31.

Kerrest, A. (2011). Outer space as international space: Lessons from Antarctica. In *Science diplomacy: Antarctica, science, and the governance of international spaces*. Washington, DC.: Smithsonian Contributions to Knowledge. https://doi.org/10.5479/si.9781935623069.133.

Kliewer, B. W. (2013). Why the civic engagement movement cannot achieve democratic and justice aims. *Michigan Journal of Community Service Learning, 19*(2), 72–79.

Launius, R. (2014). The Apollo mission and the challenge of solving the plight of the poor. *Roger Launius' Blog*. https://launiusr.wordpress.com/2014/06/06/the-apollo-11-mission-and-the-challenge-of-solving-the-plight-of-the-poor/. Accessed 20 June 2020.

Madrigal, A.C. (2012). Moondoggle: The forgotten opposition to the Apollo program. *The Atlantic*. https://www.theatlantic.com/technology/archive/2012/09/moondoggle-the-forgotten-opposition-to-the-apollo-program/262254/. Accessed 20 June 2020.

Mavelli, L. (2018). Citizenship for Sale and the neoliberal political economy of belonging. *International Studies Quarterly, 62*(3), 482–493.

McGregor, G. (2009). Educating for (whose) success? Schooling in an age of neo-liberalism. *British Journal of Sociology of Education, 30*(3), 345–358.

Morgan, P. L., Farkas, G., Hillemeier, M. M., & Maczuga, S. (2016). Science achievement gaps being very early, persist, and are largely explained by modifiable factors. *Educational Researcher, 45*(1), 18–35.

Naidoo, R. (2004). Fields and institutional strategy: Bourdieu on the relationship between higher education, inequality and society. *British Journal of Sociology of Education, 25*(4), 457–471. https://doi.org/10.1080/0142569042000236952.

National Aeronautics and Space Administration. (2020). *NASA Spinoff*. https://spinoff.nasa.gov/. Accessed 20 June 2020.

Parkinson, C., & Jones, T. (2018). Aboriginal people's aspirations and the Australian Curriculum: A critical analysis. *Educational Research for Policy and Practice, 18*(1), 75–97. https://doi.org/10.1007/s10671-018-9228-4.

Phillips, L. T., Stephens, N. M., Townsend, S. S. M., & Goudeau, S. (2020). Access is not enough: Cultural mismatch persists to limit first-generation students' opportunities for achievement throughout college. *Journal of Personality and Social Psychology*. https://doi.org/10.1037/pspi0000234.

Rai, S, & Griffiths, M. (2016). 'Useful' civic hacking for environmental sustainability: Knowledge transfer and the International Space Apps Challenge. *The Journal of Community Informatics, 12* (1). http://ci-journal.net/index.php/ciej/article/view/1124. Accessed 20 June 2020.

Resnick, B. (2020). The night sky is increasingly dystopian. *Vox*. https://www.vox.com/science-and-health/2020/1/7/21003272/space-x-starlink-astronomy-light-pollution. Accessed 10 July 2020.

Rodéhn, C. (2019). Science centres, gender and learning. *Cult Stud of Sci Educ, 14*, 157–167. https://doi.org/10.1007/s11422-018-9880-2.

Sagan, C. (1994). *Pale blue dot: A vision of the human future in space*. New York: Random House.

Scott-Heron, G. (1970). Whitey on the Moon. Small Talk at 125th and Lenox.

Senghore, F., Campos-Nanez, E., Fomin, P., & Wasek, J. S. (2015). Applying social network analysis to validate mass collaboration innovation drivers: An empirical study of NASA's International Space Apps Challenge. *Journal of Engineering and Technology Management, 37*, 21–31. https://doi.org/10.1016/j.jengtecman.2015.08.007.

Shammas, V. L., & Holen, T. B. (2019). One giant leap for capitalistkind: Private enterprise in outer space. *Palgrave Communications, 5*, 10. https://doi.org/10.1057/s41599-019-0218-9.

Simis, M. J., Madden, H., Cacciatore, M. A., & Yeo, S. K. (2016). The lure of rationality: Why does the deficit model persist in science communication? *Public Understanding of Science, 25*(4), 400–414. https://doi.org/10.1177/0963662516629749.

Sladek, M. R., Doane, L. D., Luecken, L. J., Gonzales, N. A., & Grimm, K. J. (2020). Reducing cultural mismatch: Latino students' neuroendocrine and affective stress responses following cultural diversity and inclusion reminder. *Hormones and Behavior, 120*, 104681. https://doi.org/10.1016/j.yhbeh.2020.104681.

Stephens, N. M., & Townsend, S. S. M. (2015). The norms that drive behavior. *Journal of Cross-Cultural Psychology, 46*(10), 1304–1306. https://doi.org/10.1177/0022022115600264.

Stimers, P. (2020). The President's space resources executive order: A step in the right direction. *The Space Review*. https://www.thespacereview.com/article/3925/1. Accessed 30 June 2020.

Storr, C. (2020) Could corporations control territory in space? Under new US rules, it might be possible. *The Conversation*. https://theconversation.com/could-corporations-control-territory-in-space-under-new-us-rules-it-might-be-possible-138939. Accessed 30 June 2020.

Timms, M. J., Moyle, K., Weldon, P. R., & Mitchell, P. (2018). *Challenges in STEM learning in Australian schools: Literature and policy review*. Melbourne: Australian Council for Educational Research (ACER). https://research.acer.edu.au/policy_analysis_misc/28. Accessed 15 July.

Trench, B. (2008). Towards an analytical framework of science communication models. In *Communicating science in social contexts* (pp. 119–135). Dordrecht: Springer.

UTS. (2019). *Students look to new frontiers*. https://www.uts.edu.au/news/tech-design/students-look-new-frontiers. Accessed 15 July 2020.

Voskia, A. (2020). The ecological significance of the overview effect: Environmental attitudes and behaviours in astronauts. *Journal of Environmental Psychology, 70*, 101454. https://doi.org/10.1016/j.jenvp.2020.101454.

Vromen, A. (2003). 'People try to put us down ...': Participatory citizenship of 'Generation X'. *Australian Journal of Political Science, 38*(1), 79–99. https://doi.org/10.1080/1036114032000056260.

Chapter 3
The Next Frontier of the Global Commons

Kathryn Robison Hasani

Abstract As nations and commercial entities expand further and more permanently into outer space, governance of that realm as a global commons becomes both more important and more difficult. This chapter addresses the question of what the future of space exploration and exploitation will look like as technology continues to outpace regulation and governance and more nations and actors gain access to its resources. How will nations function under old treaties and norms when faced with technical challenges or the acquisition of resources brought on by space debris or asteroid mining? The past governance of other areas of the global commons, especially the oceans, provide insight into the former question, while the history of the developed world's territorial expansion sheds light on the latter. This chapter asks if the future will resemble the utopian ideals of Star Trek or will it devolve into a Star Wars-esque past of colonialism and profiteering and begins the work of finding an answer.

3.1 Introduction

The 2020 Presidential Election in the United States, as of summer 2020, is being framed in terms of the hit movie franchise Star Wars. Donald Trump, the incumbent, has characterized himself as the Empire, ready to do battle against the Democrats with his Death Star (Robertson 2020). Joe Biden, the presumptive Democratic nominee, has been cast in the role of the Rebel Alliance, underfunded and forced into hiding, but with a righteous cause. As the governance of this area of the global commons is heavily influenced by the United States, which has historically functioned as a major player in the setting of international space policies and exploration agendas, it is only appropriate to begin with such a metaphor (Weldes 2003). Many people, when asked about outer space, turn to the science fiction

K. Robison Hasani (✉)
The University of Alabama, Tuscaloosa, AL, USA
e-mail: krobison@crimson.ua.edu

M. de Zwart, S. Henderson (eds.), *Commercial and Military Uses of Outer Space*, Issues in Space, https://doi.org/10.1007/978-981-15-8924-9_3

franchises which have shaped many of our collective understanding about space and its use even before we were able to understand what politics were and how it shapes our lives (Campbell and Gokcek 2019). In addressing the future of space as a global commons, we think of two divergent paths – one which leads us into the past and the other into the future. Will our continued push into outer space as a species resemble Star Wars or shall it be Star Trek?

Global commons refer to the natural resources and domains which are generally considered either public/common goods or common-pool resources such as the oceans, the polar regions, and outer space. Global governance of these commons is necessary to protect the "unimpeded access [which] is paramount to the progress, well-being, and stability of the entire international community" (Jasper 2012). From the very beginning of our forays into the next frontier of the global commons, like the forays into the commons of the past, war and politics have led the way. Humanity sought to escape our atmosphere, not for great discovery, but to seek competitive advantage against our enemies. The Soviet Union and the United States began the space age locked into a great space race to gain the upper hand against its greatest rival (Dolman 2005). As we move more permanently into this final frontier of the global commons, we must ask ourselves: Will the technology of space, seen as soft power in diplomacy, become hard as we move toward the era of the Space Force?

3.2 Where Will We Go? A Brief Overview of Recent U.S. Policy in Civilian Human Spaceflight and Exploration Goals

Traditionally, the U.S. has been a leader in setting the goals of deep space exploration, such as pushing to land humans on Mars by the 2030s (Grimard 2012). However, the Trump Administration has recently refocused its attention back to a return to the Moon, following the lead of other nations and space agencies' policy shifts in this direction. Notably, Jan Wörner, the Director General of the European Space Agency (ESA), has advocated for the concept of a "Moon Village" since 2015. This village, so named because it implies a community working together without first determining the exact nature of its partnership, is meant to serve as a permanent presence of humanity on the Moon to enable both scientific research and commercial ventures (Wörner 2016). Perhaps as part of his "America First" agenda, President Trump and his administration have followed the international lead toward revisiting the Moon and is determined to be the first to land upon its surface again (Tronchetti and Liu 2018).

Prior to the current U.S. administration, the National Space Policy (NSP) of the United States, last updated by President Barack Obama in 2010, precluded a return to the Moon by NASA astronauts and instead focused on a human mission to Mars, as outlined by the civilian space agency in its Journey to Mars (2014) plan. While many within the space community advocated for a return to the Moon as a testbed for the human exploration of Mars, the Obama Administration adopted a "been there, done that" attitude toward future exploration of the surface of the Moon and preferred to leave it to their international and commercial partners (Lele 2010). Instead, NASA planned only for a return to cis-lunar space and a construction of a Deep Space Gateway with commercial partners.

This itself was a departure from some of the policies of the administration of President George W. Bush who, in planning for the retirement of the Shuttle Program as a part of his 2004 Vision for Space Exploration, oriented human exploration in the U.S. toward establishing an exploration and research program on the Moon, including both robotic and human missions of extended durations with an eye toward future missions to other destinations, starting with Mars (Tatarewicz 2009; National Aeronautics and Space Administration. n.d.). The Bush Administration also launched Constellation, a crewed program including new launch vehicles and spacecraft which was meant to carry humans toward Bush's vision of America's future in space. This program, after a review initiated by the Obama Administration found that the program's budget was unsustainable, was defunded by President Obama in the 2011 federal budget (Lambright 2019; Review of US Human Spaceflight Plans Committee 2009). However, parts of the Constellation Program are still in use today – the Orion spacecraft and the Space Launch System (SLS), a modified version of the super heavy-lift expendable launch vehicle, Ares V, which never left the drawing board and, unlike SLS, did not have the capacity to launch Orion (Lunar and Planetary Institute 2019).

Both of these Constellation leftovers, along with elements of the Obama Administration's Asteroid Redirect Mission, have been folded into the Trump Administration's exploration program, Artemis, launched in 2019 (von Ehrenfried 2020). Also, like its predecessor, the Journey to Mars, Artemis includes both international and commercial partnerships. These partnerships do more to enable broad participation in the exploration of space and the potential exploitation of its resources. However, while the programs such as the Commercial Crew and Cargo Program (C3PO), launched as part of the American Recovery and Reinvestment Act of 2009, have both decreased costs and increased access, space travel, especially launches, remains prohibitively expensive for many nations and companies. To ensure equitable access to space and its resources, as enshrined in the Outer Space Treaty (OST) of 1967, nations must work together.

3.3 International Cooperation in Space

Space is inherently international, yet uniquely nationalistic. The governance of this area of the global commons is heavily influenced by the U.S. due to the size and reach of its space programs. As more nations pursue space exploration programs, the list of major players in this area has expanded to include Europe, Russia, China, and India. Political communication between these players contributes to the use and exploitation of this area of the global commons, including how the U.S. works with its international allies to govern space. As nations continue to push back against the constraints of the Outer Space Treaty of 1967, especially as it concerns the peaceful uses of outer space, the governance of this area has the power to change the dynamics of foreign policy.

In 1998, Sadeh et al. developed a theoretical framework to model international cooperation which describes the roles of policy preferences and processes of the institutions as well as the role of epistemic communities in the formation of policies and agreements for international cooperation at the level of the state actor (national space agencies). Though the state, as a unit of analysis is necessary, it may not be sufficient; future work may need to consider actors both above and below the state level, especially as more commercial actors become involved. However, this framework is important to understand what factors may be influencing statements made by heads of agencies or policies set by these agencies for international cooperation.

Broniatowski et al. (2008) build on the idea of the need for a framework for evaluating international cooperation that specifically addresses concerns from both the political and technical aspects of proposed cooperation; they also recognize that existing frameworks do not address the complexity of issues that set space policy in the United States, and the reality of political concerns that shape the way in which international cooperation works in space exploration. Building upon the literature above, developing a more holistic framework for international cooperation in space may be useful in explaining the behavior of states in space.

Weppler (2014) moves the focus away from the general topic of human exploration to the specific accomplishments in international cooperation on board the International Space Station (ISS) and highlights how these advances, including environmental control systems, contribute toward international cooperation in human exploration beyond low Earth orbit (LEO).

Technological accomplishments are also important incentives for international cooperation. Laurini and Gerstenmaier (2014) discuss how the Global Exploration Roadmap outlined ways in which space-faring agencies have come together to plan for future exploration. They offer suggestions for how those plans can be made more concrete and how partnerships to achieve those plans can be solidified, all of which are dependent on how states decide to interact with one another. Finally, Borowitz and Battat (2016) write about the next steps needed to reach the surface of Mars, though their claim regarding consensus about the surface of Mars may face more tension than they admit. However, they do bring up valuable points about how the lessons learned from missions to cis-lunar space, or an asteroid redirect mission, are critical to the development of any further deep space exploration missions, especially those to Mars.

3.4 Regulation and Governance

Beyond international cooperation in space exploration goals, scholarship in this area is also concerned with how access and use of space is governed. Larsen (2019) offers several visions for the future of space, based on which order of governance is follows. If we accept the premise of space as a global commons (as this chapter does), then we must look to the lessons of economist Elinor Ostrom, who outlines management principles for global commons which rely on regulations, monitoring, and sanctions as tools to ensure the commons is protected and accessible for all (Larsen 2019). A challenge to this premise, raised by Larsen, is whether this type of management can align with others who may see outer space as a commercial or military domain. In these conversations, the rhetoric of space for "the benefit of all" (OST) and toward the benefit of those with the ability to access or dominate it.

The space age was born out of political strife and the desire to dominate, and while obligations under the OST prevent the use of space for non-peaceful purposes, nations "retain the right of self-defense in outer space" (Larsen 2019). National space policy in the U.S. and other nations has long considered defense-related activities to as peaceful uses of outer space. However, policy shifts which began under the Obama Administration in the National Defense Authorization Act of 2015 have been exploited by the Trump Administration. President Trump, perhaps due to his hunger for military pomp and circumstance, oversaw the creation of the United States Space Force (USSF), the sixth branch of the U.S. military whose mission is to "organize, train, and equip space forces in order to protect U.S. and allied interests in space and to provide space capabilities to the joint force" (USSF 2020). The creation of the USSF provided little new capabilities, but rather largely rebranded the existing Air Force Space Command. Harrison (2018) explains that the creation of a Department of Space, mentioned as early as 2001 in government space policy reports, would have been a preferable first step rather than the creation of a new military branch, but that measured approached with multiple steps are difficult to achieve in Washington, D.C.

The USSF firmly orients the U.S. position toward military space as one which must be dominated. While military dominance in space is a concern and possible threat to its status as a global commons, seeing it as such may actually protect that definition. If, as posited by Deschenes (2019), the U.S. dominates space by treating that domain in the same way it handles the high seas, then it will also respect its similarities and accepts it as a global commons. Interestingly, according to Chinese scholarship on the global commons of the seas and outer space, China also seems predisposed to treat the latter in this way, despite its more aggressive stance on the sea (Freeman 2020). Steer (2017) argues that nations must practice restraint and need to come together collaboratively to build a modern framework for regulation and operation in outer space. In recent years, efforts within the United Nations to draft a treaty which would address the use of weapons in space has seen China and Russia as allies against their Western counterparts (Steer 2017).

These concerns have become more pressing as all users of space grapple with the increasing issue of space debris and crowded orbits. Warfare in space, especially that which would see the destruction of enemy satellites, has the potential to greatly increase space debris, and perhaps even causing space to become inaccessible as a result of the Kessler Effect (Larsen 2019). This effect, wherein the debris field around Earth becomes so congested after cascading collisions it may be impossible to safely launch into space from Earth, is one which anti-satellite weapons (ASAT) tests have magnified, especially that of China in 2007, which resulted in thousands of pieces of debris large enough to be trackable (Weeden 2007). In addition, collisions of objects in space, such as the 2009 collision of a defunct Russian satellite with an active U.S. communications satellite, are an additional risk. The aforementioned collision between two satellites, which had never before happened in space, also resulted in thousands of trackable pieces of debris (Weeden 2009). As space becomes more accessible, and with the advent of satellite constellations, such as those operated by SpaceX and Planet Labs, the problem is only getting bigger. With the danger of space debris ever present, there is perhaps no issue more important to all users of space, whether they be civil, military, or commercial, and it is the one which may be most likely to result in international agreement on the regulation of space (Weeden and Samson 2019).

However, it can be difficult to even engage in a discourse around the nature of outer space as a global commons, as not everyone agrees how it should be treated, what a commons is, or even if it is one. According to Freeman (2020), Chinese scholars argue that the use of the terminology of the global common is a "tool of 'discourse power'" intentionally used by Western space powers to restrict the influence of emerging space powers. Tepper (2019) argues that the term "commons" is problematic, not because outer space is not a global commons, but that the legal and economic definitions are often conflated and used interchangeably. He points to the need to fully understand the differences of the economic term and the legal regime in order to properly design and understand the governance of space (Tepper 2019). Despite the disagreement around the definition of global commons, it can be said that most scholars, and even perhaps most states, accept outer space as a global common.

The same cannot be said for all states. While some U.S. officials have made public statements in support of space as a global commons, some prominent figures have not. Scott Pace, the Executive Secretary of the National Space Council (revived by the Trump Administration), expressly rejected the idea of space as a global common or as a public good during remarks given in 2017. This rejection is unsurprising given the interests of the U.S. in protecting private property rights in space, especially in regard to mineral resources. The Commercial Space Launch Competitiveness Act of 2015 which protects the rights of U.S. citizens to exploit asteroid and space resources without interference from the federal government; Luxembourg and the United Arab Emirates, with the Space Resources Act of 2017 and the Federal Law No. 12 of 2019 on the Regulation of the Space Sector, respectively, have passed similar legislation. The ways in which space resources are treated and governed by national and international regulations will have significant impacts on the future of access and use of space (De Man 2017).

3.5 Space Resources and the Democratic Peace

The potential value of asteroid or other space resources cannot be understated. To consider what effect these will have on outer space as a global commons, it is helpful to turn to political science for clues. Power and resources play a significant role in international relations; those states which have both power and resources are able to maintain their internal and external security more easily than those without either or both. However, as technology continues to increase at a rapid pace, what will happen once states begin to utilize the resources of space? More specifically, what are the implications to democratic peace from territorial claims in space? What are the implications to the world economic system from space resource mining? Though these are two distinct questions, they are interrelated and therefore it is useful to consider them together. To consider these questions, the prohibition against making territorial claims in space by the Outer Space Treaty of 1967 must be set aside, as there are compelling arguments to assume this may be likely to occur in the future (Buxton 2004; Coffey 2009).

There are two relevant theories from within the discipline of International Relations, Realism and International Political Economy (IPE). Realism (and its critiques) is useful to address these questions because Realism is about power, and the act of claiming territory is an act of power. Morgenthau (1985) argues that all politics is motivated by self-interest and Waltz (1979) argues that the international system changes when power changes. These arguments are relevant to the question of territorial claims in space because a nation able to claim and hold territory in space will achieve an amount of power yet unknown and possibly not measurable at this time. This is why the second question of the implications of space resources must also be considered in conjunction with the first because it is helpful to assume that a nation makes a territorial claim in space to protect its interest in resources in that territory.

This is where it becomes necessary to utilize IPE theories. Gilpin (1981) explores both politics and economics and considers how these lead to an unequal growth of power among different states. Gilpin (2016) considers the impact of technological change on world systems of politics and economy and argues for a "benign mercantilism" (p. 408) as necessary for global security. Benign mercantilism is subordinate to politics (Gilpin 2016), and if we are to explore the implications of bringing resources back to Earth from space, worth possibly trillions of dollars, unless there is some political control over the economic systems of the planet, it risks ruining the economies of nations unable to access space resources. Although the circumstances which would allow the situations posited above have not yet occurred, the technology to make both feasible is currently available. However, since no nation has yet made a territorial claim in space, and no one has successfully mined an asteroid for resources to return to Earth, we cannot study these questions directly at this time. However, we can make inferences based on historical situations which dealt with incredibly similar circumstances in order to make informed and educated inferences about the future actions of both states and non-state (likely commercial entities) in reference to the questions above.

What is the democratic peace, and why is it important to consider it when imagining the next frontier of space? The democratic peace is an idea first put forward by Kant in 1796 and that is still being refined by political scientists today (Russett 1994; Rosseau et al. 1996; Hegre et al. 2020) that argues that democratic nations are more likely to maintain peace with each other than to be in conflict. It is not a question of *if* the democratic peace will be affected by territorial claims in space, but rather a question of *how*. The returning of space resources to Earth perhaps highlights the question of what the future of space will look like more than any other. Will states hoard and ration space resources, treating the next frontier as the colonizing force of the Empire in Star Wars, or will they seek to redistribute the wealth to the benefit of all as the Federation does in Star Trek?

If territorial claims are made in space it will lead to a shifting of the balance of power among nations toward nations able to make those claims. However, it is unlikely that these territorial claims would lead to conflict, but rather that these claims would lead to increased international agreements and partnerships. The nations who are most able to make these claims and return resources to Earth are those which already hold a disproportionate amount of global wealth today. Therefore, it is unlikely that those nations which already lack access to space and its technologies would be in a position to pursue conflict with those states or actors capable of territorial claims. It is much more probable that those without access to space resources will seek to align with nations with those capabilities. This theory is informed by Samuelson's (1967) claims about specialization in goods and services and is interrelated with the how the world economy reacts when incredibly valuable resources are returned to Earth from space.

Non-territorial states, in lieu of fighting over resources, will seek to form alliances with territorial nations with the ability to obtain the resources. The most expedient way to secure alliances would be the ability to offer a valuable service or product useful to the process of obtaining the space resource, a model which is currently in practice already in space exploration (Long 2016). Again, this assumes that nations will make territorial claims in space because of the resources available on that planet or asteroid. To be sure, an argument could be made that if the above hypotheses are found to be correct, and if two or more nations obtained the ability make territorial claims in space and ability to exploit the resources in those territories then conflict between those nations is likely (Beard 1934; Carr 1939; Booth 2007). However, that possibility, while important, is mostly beyond the purview of this chapter. Rather, this chapter is concerned with how defining space as a global commons shapes its future utilizations.

While setting aside the prohibition against territorial claims was useful to the discussion above, it is not necessary. Tepper (2019) points out that using the framework of an economic commons to define some parts of space (such as an asteroid) allows the extraction of resources without the need for territorial claims. Instead, the actors capable of territorial claims instead become authorized users able to assert property rights over any resources they extract from the economic commons and return to Earth. In all likelihood, the first users who accomplish this will not be state actors, but rather corporations. In recognition of this, we must consider how commercial users of space are governed and what obligations they have under the OST.

3.6 Commercial Use and Exploitation of Space

Almost from the very beginning of space, commercial companies have been involved in its use, exploration, and exploitation. Governments around the world, including in the U.S. partnered with private companies in both civil and military ventures. The first commercial space transportation company was Arianespace, based in Europe and still operating today, though not without stiff competition from SpaceX (Lal 2013). The U.S., through C3PO and other programs, has transferred the bulk of its launch activities to the private sector, including all of its LEO activities. With SpaceX's successful Demo-2 launch of Crew Dragon in May 2020, this also includes crew transportation; NASA has purchased just one additional Soyuz seat from Russia for 2020 and intends rely on its commercial providers for transportation to the ISS going forward.

Commercial launch service providers (LSPs) are not only concerned with government contracts, though some providers derive the bulk of their income from them. We can think of LSPs in several categories: Government-oriented (GO-LSPs), Hybrid (H-LSPs), and Private (LSPs). In the U.S., the type of launch service provided affects regulation under which commercial launches operate. Civil and national security launches do not require direct government regulation because they are carried out by the U.S. government under its own oversight; commercial space launches and reentries with government payloads do not require licensing from the Federal Aviation Administration (FAA), but, with very limited exceptions, all other launches and reentries carried out by commercial companies must be licensed by the FAA (Aeronautics and Space 2006).

Regulation for commercial spaceflight is determined by the launch payload (payload refers to the type of spacecraft launching, and can include anything from satellites, cargo to the ISS, or spacecraft designed for humans) in instances of spaceflight launched from atop traditional launch vehicles (rockets). Launch vehicles can be either expendable or reusable, and also includes other launch systems such as Virgin Galactic's air launched Space Ship Two, or balloon launched systems. The FAA's Office of Commercial Space Transportation (AST) has the responsibility of regulating all non-governmental launches in the United States, along with licensing commercial LSPs' reentry (return from space). AST issues licenses for orbital flights and issues permits for suborbital flights. Orbital flights are those which have sufficient velocity to go into various orbital inclinations around the

Earth or the sun, suborbital flights are those which reach space (generally accepted as above the Kármán line at 100 kilometers above sea level), but do not have sufficient velocity to go into orbit (AST).

NASA and the Department of Defense (DOD) handle regulations for government launches. Hancock (2005) explains how the Commercial Space Launch Act (CSLA) provided for regulation of non-governmental launches through FAA oversight. The CSLA is necessary, though not mentioned by Hancock (2005), due to constraints on the United States due to its signatory status to the OST, which requires the United States to supervise all space activities carried out by non-governmental entities (Dempsey 2016). The FAA's AST provides guidelines to commercial companies in order to fulfill this obligation under the OST, as well as guidance on the correct interpretation of legislation pertaining to space law through the publication of rules which govern the commercial space industry.

Hertzfeld (2002) addresses the salience of commercial space regulation in the U.S. due to the status of the U.S. as the single largest investor in the space economy, both public and private. Examining the example of the commercial industry in the U.S. is also salient because it spends more money on space than the six largest space economies below it combined. However, Hertzfeld also points out issues that the U.S. dominance in space can cause problems for globalization and sharing of technologies, which may in turn depress commercial space ventures; these issues mainly center on concerns about and restrictions to space activities in terms of national security as discussed earlier in this chapter. However, this concern does not stop the U.S. government from pursuing policies that significantly increased commercial involvement in U.S. civil and defense space, as explored in Lambright's (2015) examination of NASA's Commercial Orbital Transportation Services (COTS) program. In these programs, including the COTS, the Cargo Resupply Services (CRS) to the ISS, and the various Commercial Crew Programs (CCP), NASA has contracted with private companies to develop and procure services to transport cargo and crew to the ISS and to LEO. The companies currently launching cargo programs are SpaceX and Orbital ATK, while both SpaceX and Boeing hold contracts for crewed (NASA Office of the Inspector General 2016).

It is clear that commercial companies, both in the U.S. and around the world, are able to access space and have been supported in their endeavors by large investments from the public sector. They are regulated by their governments and their governments are obligated to ensure they are in compliance with the OST. The question as technology improves and more commercial actors venture into space regularly and more permanently, what happens when they are the ones who bring space resources into the economy. If we look to the past, to other exploitations of the global commons, the story is only positive for a small set of actors. The legacy of colonial explorers' exploitation of newly encountered resources is still felt today in the gap between developed and developing nations. If, as Article I of the OST contends, space is for the benefit of all, and should be free and accessible for all nations, "irrespective of their degree of economic or scientific development," then the legislation which has enabled the boon of the commercial space economy may also precipitate the bust of those without the means to access space.

3.7 Equitable and Just Access to Space

The power dynamics of the post-colonial world in which we live shape almost every facet of our life, space included. If we do not take care to prevent the systems from replicating themselves, then they will become part of our space heritage, and in fact, already are part of it. Diversity is noticeably lacking in the largest space companies. While we are beginning to see a more representative generation in the space workforce, the most recent crewed spaceflight from the U.S. was undertaken by two white male astronauts and guided by a control room of mostly white, mostly male faces. This did not go unnoticed, and while NASA's next scheduled human spaceflights include women and people of color, the boards and leadership of many involved in the space industry reflect the reality of a legacy of colonialism.

The "vast disparities between developed and developing states have emerged from a history of massive and grievous injustices including colonialism and slavery which explains why the former states are affluent and the latter are poor" (Sutch and Roberts 2019). Addressing these injustices and preventing their replication is more easily managed through the management of space as a global commons than treating it as territory which may be claimed. However, how we treat that commons will determine which future we look at. If the commons is treated as *res nullius*, where anything can be owned, then our future will resemble the past. If we treat is as *res communis*, where it is open to all but owned by none, we may be able to forge a new path without repeating the mistakes of the colonial powers (Husby 1994). The legislation and policies which are currently being written and enacted will determine whether we repeat the sins of the past, and venture too close to creating a sort of Space Industry Trading Company with practices uncomfortably close to its namesake.

Centering international space governance around the Sustainable Developments Goals (SDGs) is one way in which the UN could ensure equitable access in space. The aim of the goals is to ensure peace and prosperity for all people on the planet (Morton et al. 2017). Since all UN nations have agreed to these goals as of 2015, and because of the great promise of space technology in helping to achieve these goals, they could become a framework on which to build new norms of operation for commercial actors in space, especially those who would return highly profitable resources from space. A strong governance of the global commons regime in space could mandate policies which ensure that all states are able to profit from resources common to all of humankind and ensure their equitable and just access to space and space technology.

3.8 What Does the Future Hold?

Space policy as political power is increasing more quickly than traditional scholars of political science are aware. This is due in part to several factors as discussed in this chapter, like the increasing involvement of commercial actors in space policy, but also those not discussed such as the public's inattention to space.[1] However, strong international partnerships between close political allies of the U.S. are now in the process of being made with emerging actors in the space industry, like China, without U.S. involvement. It is clear that it will not just be the U.S. and Russia who make significant impacts on the direction of space exploration going forward, that China and India, along with commercial companies are all crafting their own vision for the future of space. What we do not yet have is enough information about these newer actors in the context of political science to understand what political power these changes might yield to the states involved. As the pace of development continues to increase, and the utilization of space becomes more ubiquitous, we will begin to see the shape of the future.

In consideration of these unknowns, a new paradigm of international relations in space and the governance of it as a global commons must arise from the current state of affairs. We cannot yet predict who will emerge as the leader of the next great space race, whether that is a race to space weapons or resources, but we can trust that whichever actor emerges as the leader may gain significant political and economic power in the decades to come. Science fiction has given us two realities to consider when we imagine the future of space, the colonialism of Star Wars or the equity of Star Trek. Those who are most successful at exploiting the next frontier of space will claim the power to shape whether that future will be dominated by the thirst for dominance of the Empire or by the desire for peace and prosperity of the Federation.

References

Aeronautics and Space, 14 C.F.R. § 400–460. (2006).

Beard, C. A. (1934). *The economic basis of politics*. Transaction Publishers.

Booth, K. (2007). *Theory of world security* (Vol. 105). Cambridge University Press.

Borowitz, M., & Battat, J. (2016). Multidisciplinary evaluation of next steps for human space exploration: Technical and strategic analysis of options. *Space Policy, 35*, 33–42.

Broniatowski, D. A., Cardin, M. A., Dong, S., Hale, M. J., Jordan, N. C., Laufer, D. R., Matheiu, C., Owens, B. D., Richards, M. G., & Weigel, A. L. (2008). A framework for evaluating international cooperation in space exploration. *Space Policy, 24*(4), 181–189.

Buxton, C. R. (2004). Property in outer space: The common heritage of mankind principle vs. the first in time, first in right, rule of property. *Journal of Air Law and Commerce, 69*, 689.

Campbell, J. R., & Gokcek, G. (2019). *The final frontier: International relations and politics through star trek and star wars*. Lanham: Rowman & Littlefield.

Carr, E. H. (1939). *The twenty years' crisis, 1919–1939: An introduction to the study of international relations*. New York: St. Martin's Press.

Coffey, S. (2009). Establishing a legal framework for property rights to natural resources in outer space. *Case Western Reserve Journal of International Law, 41*, 119.

[1]Unless, of course, Elon Musk is launching a spacesuit to orbit in his Tesla.

De Man, P. (2017). *Luxembourg law on space resources rests on contentious relationship with international framework*. Leuven Centre for Global Governance Studies.
Dempsey, P. S. (2016). National laws governing commercial space activities: Legislation, regulation, & enforcement. *Northwestern Journal of International Law & Business, 36*, 1.
Deschenes, C. N. (2019). Enabling leaders to dominate the space domain. *Military Review, 99*(3), 109.
Dolman, E. C. (2005). *Astropolitik: Classical geopolitics in the space age*. Routledge.
Exploration Hardware. (2019). Lunar and Planetary Institute. https://www.lpi.usra.edu/lunar/constellation/. Accessed 21 Feb 2020.
Freeman, C. P. (2020). An uncommon approach to the global commons: Interpreting China's divergent positions on maritime and outer space governance. *The China Quarterly, 241*, 1–21.
Gilpin, R. (1981). *War and change in world politics*. Cambridge University Press.
Gilpin, R. (2016). *The political economy of international relations*. Princeton University Press.
Grimard, M. (2012). Will the US remain the real leader of human space exploration? A comparative assessment of space exploration policies. *Acta Astronautica, 75*, 1–14.
Hancock, R. (2005). Provisions of the Commercial Space Launch Act (CSLA). *Space Policy, 21*(3), 227–229.
Harrison, T. (2018). Why we need a Space Force. *Center for Strategic and International Studies (CSIS), 3*.
Hegre, H., Bernhard, M., & Teorell, J. (2020). Civil society and the democratic peace. *Journal of Conflict Resolution, 64*(1), 32–62.
Hertzfeld, H. R. (2002). Measuring the economic returns from successful NASA life sciences technology transfers. *The Journal of Technology Transfer, 27*(4), 311–320.
Husby, E. (1994). Sovereignty and property rights in outer space. *Journal of International Law and Practice, 3*, 359.
Jasper, S., & Mahnken, T.G. (2012). Conflict and cooperation in the global commons: a comprehensive approach for international security. Georgetown University Press.
Kant, I. (1796). *Project for a Perpetual Peace: A Philosophical Essay*. Vernor and Hood.
Lal, B. (2013). A Brief History of Government Policies to Promote Commercial Space. *Journal of the Washington Academy of Sciences, 99*(3), 25–41.
Lambright, W. H. (2015). Launching commercial space: NASA, cargo, and policy innovation. *Space Policy, 34*, 23–31.
Lambright, W. H. (2019). Administrative leadership and long-term technology: NASA and the international space station. *Space Policy, 47*, 85–93.
Larsen, P. B. (2019). Outer space: How shall the world's governments establish order among competing interests. *Pac. Rim L. & Pol'y J, 29*, 1.
Laurini, K. C., & Gerstenmaier, W. H. (2014). The global exploration roadmap and its significance for NASA. *Space Policy, 30*(3), 149–155.
Lele, A. (2010). The US Space Priorities under the Obama Administration. *Strategic Analysis, 35*(1), 31–35.
Long, J., (2016). China's space station project and international cooperation: Potential models of jurisdiction and selected legal issues. Space Policy, 36, pp.28–37.
Morgenthau, H. J., Thompson, K. W., & Clinton, W. D. (1985). *Politics among nations: The struggle for power and peace*. McGraw-Hill.
Morton, S., Pencheon, D., & Squires, N. (2017). Sustainable Development Goals (SDGs), and their implementation: A national global framework for health, development and equity needs a systems approach at every level. *British Medical Bulletin*, 1–10.
National Aeronautics and Space Administration. (n.d.). *Humanity's return to the moon*. Retrieved May 2, 2020, from https://www.nasa.gov/specials/artemis/
National Aeronautics and Space Administration, Office of the Inspector General. (2016). *NASA's commercial crew program: Update on development and certification efforts (report no. IG-16-028)*. https://oig.nasa.gov/docs/IG-16-028.pdf

Robertson, D. (2020, May 19). What liberals don't get about Trump supporters and pop culture. *Politico Magazine*. https://www.politico.com/news/magazine/2020/05/16/trump-death-star-pop-culture-mutiny-bounty-curb-enthusiasm-260471

Rousseau, D. L., Gelpi, C., Reiter, D., & Huth, P. K. (1996). Assessing the dyadic nature of the democratic peace, 1918–88. *American Political Science Review, 90*(3), 512–533.

Russett, B. (1994). *Grasping the democratic peace: Principles for a post-Cold War world*. Princeton University Press.

Sadeh, E., Lester, J. P., & Sadeh, W. Z. (1998). Modeling international cooperation in human space exploration for the twenty-first century. *Acta Astronautica, 43*(7–8), 427–435.

Samuelson, P. A. (1967). Summary on factor-price equalization. *International Economic Review, 8*(3), 286–295.

Review of US Human Spaceflight Plans Committee, Augustine, N. R., Austin, W. M., Bajmuk, B. I., Chiao, L., Chyba, Crawley, E.F., Greason, J.K., Kennel, C.F., Lyles, L.L. and Ride, S.K. (2009). *Seeking a human spaceflight program worthy of a great nation*. National Aeronautics and Space Administration.

Steer, C. (2017). Global commons, cosmic commons: Implications of military and security uses of outer space. *Georgetown Journal of International Affairs, 18*, 9.

Sutch, P., & Roberts, P. (2019). Outer space and neo-colonial injustice. *International Journal of Social Economics, 46*, 1291–1304.

Tatarewicz, J. N. (2009). The "Vision for Space Exploration" of President George W. Bush, space science, and US space policy. *Futures, 41*(8), 531–540.

Tepper, E. (2019). Structuring the discourse on the exploitation of space resources: Between economic and legal commons. *Space Policy, 49*, 101–290.

The American Recovery and Reinvestment Act of 2009. P.L. 111-5. Stat. 131. (2009).

Treaty on Principles Governing the Activities of States in the Exploration and Use of Outer Space, including the Moon and Other Celestial Bodies, RES 2222 (XXI), (1967).

Tronchetti, F., & Liu, H. (2018). The Trump administration and outer space: Promoting US leadership or heading towards isolation?. *Australian Journal of International Affairs, 72*(5), pp.418–432.

Trump, D. J. (2017). Presidential memorandum on reinvigorating America's human space exploration program. 11 December 2017. *WhiteHouse.gov*.

U.S. Commercial Space Launch Competitiveness Act. *P.L.* 114–90. (2015).

United States Space Force (2020). United States Space Force Mission. https://www.spaceforce.mil/About-Us/About-Space-Force/Mission/

von Ehrenfried, M. (2020). The Artemis Lunar program overview. In: *The Artemis Lunar program. Springer Praxis Books* Springer, Cham. https://doi.org/10.1007/978-3-030-38513-2_2.

Waltz, K. N. (1979). *Theory of international politics*. Waveland Press

Weeden, B. (2007). Chinese anti-satellite test fact sheet. *Secure World Foundation*.

Weeden, B. (2009). Iridium-Cosmos collision fact sheet. *Space World Foundation, November, 10*, 2010.

Weeden, B., & Samson, V. (2019). India's ASAT test is a wake-up call for norms of behavior in space. *Space News*.

Weldes, J. (2003). Popular culture, science fiction, and world politics. In *To seek out new worlds* (pp. 1–27). New York: Palgrave Macmillan.

Weppler, J. (2014). Contributions of the International Space Station towards future exploration missions. *Acta Astronautica, 104*(2), 552–557.

Woerner, J. (2016). Moon Village: A vision for global cooperation and Space 4.0. *Jan Woerner's blog, ESA*.

Chapter 4
Women in Line: Space Security in the United States

Elsbeth J. Magilton

Abstract Nation-wide data on women in national security, partnered with individual interviews, paint a clear picture: the United States is a long way off from gender equality in national and space security positions. However, universities are poised to change that now more than ever before. This article gives a brief overview of the "business case for diversity" research trend showing the economic advantages of gender-diverse leadership and discusses women-focused professional organizations and female student interest in national and space security. This is followed by an introduction into U.S. space security policy and the roles women play in that sector's development. The final section highlights efforts at the University of Nebraska and offers guidance to university educators on pedagogical approaches to support gender diversity in national space security.

4.1 Introduction

There is a long-standing joke between women who regularly attend military and national security conferences: at least there is never a line for the bathroom. A friendly woman recently made this observation to me in what was a remarkably short bathroom line during a break at a major annual conference. When I asked what she did, she explained she was the wife of one of the officers in attendance. I do not diminish the sacrifice of a military spouse, including their own careers, but I must profess some disappointment that as it turned out several the women in line with me were attending as spouses rather than in a professional capacity. So, while it is a true joy to walk by a 20-person deep line for the men's bathroom and easily breeze into the women's, it does not remedy the many complex idiosyncrasies of being a woman in a room where 90% of the occupants are men. It is important to point out that this complexity is further exaggerated for women of color and for

E. J. Magilton (✉)
University of Nebraska, Lincoln, NE, USA
e-mail: elsbeth@unl.edu

M. de Zwart, S. Henderson (eds.), *Commercial and Military Uses of Outer Space*, Issues in Space, https://doi.org/10.1007/978-981-15-8924-9_4

those who are female-identifying or nonbinary. This chapter defines "woman" as anyone who identifies as such, but I encourage readers to think more deeply into diversity and investigate the research in that area.[1]

I am the executive director of several academic programs and centers, working in national security by way of my own research and the goals of my students. Much of my work intersects with security and space policy, however my most important role (or at least my favorite) is that of a mentor to individuals hoping to enter the national security and space sectors. Over the past nine years I have observed an increasing number of female students pursuing national security positions, often in connection to the space industry and outer space security. In addition to growing student interest, there is an increasing number of professional groups and organizations focused on fostering women's interest in the field. These groups benefit from a decade of research on the positive impacts of diversity to decision making. From professional networking organizations like 'Women in International Security' to the social impact business '#NatSecGirlSquad,' there has never been more vocal demand to diversify the field.[2]

My specific interest is not just how we attract female students to security policy and legal issues in outer space, but how we can train them to ensure they find jobs and advance in those positions. At the University of Nebraska, the Space, Cyber, and Telecommunications Law degree program is uniquely specialized in this field. It has over eleven years of experience preparing students to work in space and security law. There is history to look back on to help us frame our future, but first we must define our present.

This chapter examines the "business case for diversity" research movement, looks at the experiences of women in national security positions, provides an overview of U.S. space policy and women's history in military space, and finally offers pedological advice to educators seeking to help inspire and support female success in national and space security positions.

[1]There is much research available on the bias and challenges faced by women of color, and nonbinary or transgender service members. One such paper this author recommends is Goodwin et al. 2019.

[2]A list of other notable professional groups focusing on advancing diversity in foreign policy, national security, and defense: Center For Feminist Foreign Policy, CNAS Women In International Security, CSIS Diversity And Inclusion Leadership, CSIS Smart Women Smart Power, Foreign Policy Interrupted, Gender Champions In Nuclear Policy, Girlsecurity, Leadership Council Of Women In National Security, Out In National Security, Service Women's Action Network, United States Of Women, Women Of Color Advancing Peace Security And Conflict Resolution, Women's Congressional Staff Association, Women Also Know.

4.2 Education Programs, Industry Positions, and Making "The Business Case"

Having women engaged in strategic discussions, at every level, has a positive impact on outcomes – and that fact is old news. It has long been conjectured that a diversity of viewpoints leads to better decision making and more positive outcomes. In the private sector, efforts began in the 1990s to show that including women in leadership roles not only was a morally sound practice, but also provided a competitive advantage and improved performance. These efforts drove the creation of the so-called "Business Case for Diversity" research movement, which continues today (Offerann and Foley 2020).

In 2004 one such study by Catalyst, an organization devoted to expanding opportunities for women in business, examined public data from 353 Fortune 500 companies (Catalyst Staff Advisors 2004). Their findings showed that companies with the highest female representation on their top management teams showed 35% higher rates of return on equity and a 34% higher total return to shareholders (Catalyst Staff Advisors 2004). While this report did not directly show causation between women and a competitive advantage, it seemed to indicate a possible correlation. Other more current studies continue to indicate that companies with higher proportions of women in upper level management seem to achieve more profits (Lublin 2016). Outside of examining competitive performance, other research indicates that female representation on boards increases company engagement in socially responsible business methods and that those companies enjoy a better social reputation overall (Byron and Post 2016). In sum, there are significant studies that show having women in leadership roles may help business growth and company reputation in the private sector (Noland et al. 2016).

Turning to the military and defense contractors, similar "business case" research is much sparser. One limiting factor for researchers attempting these projects may be that many in the military seem to be resistant to pleas for "diversity and inclusion." This is often attributed to personnel focusing entirely on mission-effectiveness above any other concerns. Thankfully, however, there is still *some* data to speak to this (CNAS 2016). The Center for New American Security released a report in February 2017 discussing tactics and reasoning for expanding women's roles in higher level national security positions (Kidder et al. 2017).

The report thoughtfully lays out some of the historical barriers that have led to fewer women holding prominent offices due to a smaller amount of "on-ramps" in to the field (Kidder et al. 2017). Veteran status and military experience are significant factors in entry to the field of national security. These on-ramps, or career and educational pathways to institutions and positions, are considerably impacted by the fact that military academies only started accepting women in the past 40–50 years. This has delayed modern senior-level representation and academies are still making up time in achieving gender equality in their classes. Take for example, West Point Academy. A degree from West Point is a significant on-ramp into national security positions as a civilian or military member. Since 2000, the percentage of women per class varies between just 16–20% (Kidder et al. 2017).

Magnifying the low rate of female students in military institutions, is the fact that at non-military academic institutions the number of women receiving an advanced education is actually *rising*. In 2016, 57% of undergraduate degrees in the U.S. were awarded to women, and it is projected that women will continue to outpace men (Duffin 2020). In addition, female representation is also increasing at graduate schools. Graduate level degrees (whether law, business, or doctoral) in security studies, public policy, and national security are a prime talent pool for entry and mid-level employees in the national security sector. Having more women in those programs should correlate to career placements (Kidder et al. 2017). But, female enrollment still vastly outpaces the rate of women seen at the top of the sector professionally,

> The George Washington University Elliott School of International Affairs Master of Arts, Master of International Policy and Practice, and Master of International Studies program has the highest rate of women, as measured by their fall 2015 incoming class at 60 percent. The Harvard Kennedy School Master of Public Administration and Mid-Career Master in Public Administration programs and the Johns Hopkins School of Advanced International Studies Master of International Public Policy program have the lowest rates (Harvard Kennedy School as measured by five-year average, and SAIS as measured by the fall 2015 incoming class) at 35 percent. However, even programs with more modest enrollment still outpace the rate of women who make it to the top in the national security field (nowhere near 35 percent) (Duffin 2020).

This indicates that the interest is there from female students and that there is a talent pool to draw from – but even after educational achievements, women are not making it to the top of their field in the same numbers as their male peers.

A 2020 survey titled "What It Looks Like vs. What It Is: Building Competent Diversity in National Security" was a collaborative effort between Guidehouse and #NatSecGirlSquad to survey individuals working in national security (contractors, government, or military). Their results showed that 56% of survey takers strongly disagreed or disagreed with the statement that "national security careers offer a welcoming environment for women." When those results were filtered by gender, more women (60%) than men (42%) disagreed with the statement. Even more troubling, 59% of the female respondents reported that they had experienced discrimination in their role as a leader, while only 16% men said the same. 54% of women reported that they had to work twice as hard to earn their position compared to others, whereas only 12% of men reported the same (#NatSecGirlSquadxGuidehouse 2020). To say the least, these results are disappointing.

Additionally, looking solely at the employment percentages in the field does not give a complete picture. Take for example the Central Intelligence Agency (CIA). In 1983 women accounted for 37% of the total staff, but only 23% of the professional population. The majority were in clerical staff positions—figures that had remained virtually the same since 1953 (McMahon 1983). Since the 1980's, however, women's professional representation at the CIA has grown to 46% in 2012 (Kidder et al. 2017). Even more positively, in 2013 the CIA had a very visible success story for women in intelligence, sharing the role that female analysts played in the capture of Osama bin Laden. The team inspired headlines like "Hunting Osama bin Laden was Women's Work" and garnered the CIA international attention in discussing the importance of diversity on teams (Windrem 2013). The visibility of female success stories, and seeing women in these positions, helps students and young women visualize their own potential careers.

When examining the visibility of women in national security, political appointment positions are typically the most publicly recognizable being cabinet level and confirmed by the U.S. Senate. In terms of female visibility, it may be the private sector has further to go:

> In 2015, 2 of the 5 Under Secretaries of Defense, 1 of the 5 Principal Under Secretaries of Defense, and 2 of the 14 Assistant Secretaries of Defense were women (CNN Editors 2013). … Among all U.S. military generals and admirals, 7.1 percent are women. As a point of comparison, 5.4 percent of Fortune 500 (27 of 500) companies are led by female CEOs (Fairchild 2014).

So, while women have a long way to go in equal representation in national security, there is light at the end of tunnel and a growing contingency of professionals speaking up.

This section could continue to cover the social and institutional barriers women face in utilizing institutional on-ramps to national security positions and advancing through the ranks, but that is an entire book onto itself. Rather, this section was meant to introduce the concept of the "business case for diversity" and the challenges women face advancing through the national security workface. In all, these studies and numbers, while troubling, bring a growing sense of optimism when compared to even just 50 years ago.

The question you may be asking though, is "how exactly does this relate to space?"

4.3 Space Security: A Brief Look at the United States History of Space and Security

The core organizational theme of this book is, of course, space. Without a doubt, national security considerations are fundamental to any international discussions regarding the use of outer space. This section outlines U.S. history of national security activities in space, the significant role national security leaders have in space use and exploration, and prefaces a discussion on women in the space field.

Since the United States has been active in space, i.e. the beginning of the space age, national security policies and military strategy have been intertwined with space and paramount to decision making. In 1958, the U.S. National Aeronautics and Space Act passed by Congress, specified that U.S. military space activities would be conducted by the Department of Defense (DoD), while also creating the National Aeronautics and Space Administration (NASA) to conduct the nation's civil space program. Many draw a bright line between scientific space exploration and military space activities but the two have had a long, if not tenuous, relationship since the beginning.

Stretching beyond its creation in 1958, NASA has been a globally celebrated agency dedicated to science and technology related to air and space (NASA 2018). The agency, spread over nine centers, enjoys a reputation of scientific excellence and achievements in space exploration. The public tends to not associate NASA with national security issues. However, the U.S. military – particularly the Air Force – cooperates with NASA to a great extent. For example, NASA's Kennedy Space Center in Florida and the Air Force's Cape Canaveral Air Force Station are closely coordinated, sharing many facilities. To capitalize on the existing coordination, many contend there are more opportunities for NASA to participate in national security: from placing greater weight on international cooperation, to combining missions with the military, to using technological advancements to support security (Oleson et al. 2012).

In truth, whether NASA – or any private space company – wants to be involved in national security or not, national security leaders are interested in them. Think about almost any space technology; a satellite that has a camera focused on the Earth, a communications satellite, any kind of space vehicle. All of these can be used for peaceful scientific endeavors, but all can just as easily be used for intelligence gathering, surveillance, or causing outright harm to the space technology of another country. To this end, the U.S. military has been interested in space since the Soviet satellite Sputnik spurred NASA's creation over 50 years ago (McMahon 1983).

In 1988, U.S. Air Force Lt. Col. David Lupton defined four schools of thought on space power: Space Control, The High Ground, Survivability, and Space Sanctuary, all of which have enjoyed varying levels of prominence throughout U.S. space history (Lupton 1998). 'Space Control' places a focus on using space to enhance military power. Similarly, 'The High Ground' seeks to use space for projecting force and dominating engagements on Earth (for example, space-to-earth weapons or space-based missile defense). The least popular 'Survivability' school of thought focuses on minimizing reliance on space under the belief that it is a fundamentally vulnerable domain (Weeden 2019). By far the most historically prevalent, the concept of 'Space Sanctuary,' focuses on using space to acquire information for strategic decision makers. This includes, for example, using space assets (satellites) for intelligence gathering, warning systems, and treaty verification. This school of thought has largely dominated both national and international thinking on space security since the 1950's (Weeden 2019).

The U.S. has a lengthy history of including space in military decision making. Jumping to 1982, the *Air Force* Space Command was established to provide space operation support during the Cold War and Desert Storm and has continued to provide mission support and military expertise in space operations (Air Force Space Command n.d.). Distinct from the Air Force Space Command was the first U.S. Space Command (USSPACECOM), a unified combatant command.[3] USSPACECOM was established in 1985 but combined with U.S. Strategic Command in 2002 after the September 11th attacks in New York City (Erwin 2019). Since then, the U.S. has made its renewed interest in space dominance known whether it is (arguably) from a Sanctuary, High Ground, or Control approach. Recently re-activated, USSPACECOM is the newest of the eleven unified commands in the DoD (USSPACECOM 2020). As part of the Command's vision and the Commander's narrative, USSPACECOM will treat outer space as a theater of war (Erwin 2019).

The re-activation of the Command is not the only recent U.S. move in displaying space dominance. World-wide, President Donald Trump's announcement of his desire to create a "U.S. Space Force" caused a stir, though the notion of such an organization was not entirely unheard of when the President surprised the nation and the world with his announcement in 2018 (Smith 2018). Prior to that, in 2017, the U.S. House of Representatives wanted to create a Space Corps to be housed within the Air Force; analogous to the Marine Corps within the Department of the Navy. The Senate, however, wanted to create a Chief Information Warfare Officer who would oversee space, cyberspace, and information activities for the DoD. While Congress agreed to studies of both options, the Trump administration made it clear it sought a new branch of the U.S. military – an action that required Congressional authorization. While the politics continued to unfold Trump, as discussed earlier, re-activated USSPACECOM – which did not require the same congressional authorization (Smith n.d.).

In February 2019, Trump signed Space Policy Directive-4 directing his DoD to submit a proposal to Congress to create the Space Force. Eventually, however, the DoD and the Trump administration decided to forego the concept of the Space Force as new military *branch*. Instead they proposed that it be a sixth military *service* under the Air Force, similarly to the Marine Corps relationship to the Navy. If this sounds familiar, it is because this is what the House passed in 2017, but now it is the Space Force rather than the Space Corps. In the end, while the administration did not get another branch, the U.S. now has six military services (Army, Navy, Marine Corps, Air Force, Coast Guard and now Space Force) (Smith n.d.).

The Trump administration also re-created the National Space Council and has made several agency changes aimed at streamlining the regulatory hurdles space companies face (Office of Space Commerce n.d.). The administration has largely left the 2010 Obama era National Space Policy and the 2011 DoD National Security Space Strategy (meant to implement the Obama policy directive) unchanged, except for a few points – though officials say these changes "effectively replace" the 2010 policy (Smith n.d.).

---

[3] The Defense Department has 11 combatant commands, each with a geographic or functional mission that provides command and control of military forces in peace and war. A unified command has a broad and ongoing mission under a single commander that is composed of at least two service branches. See: https://www.defense.gov/Our-Story/Combatant-Commands/

In sum, this brief overview of U.S. space security is meant to show that outer space and its relationship to national security are in the public spotlight. Global leaders have a major impact on how the world sees space. From modern celebrations of "space nerds" to a President who wanted to create a military branch (incidentally inspiring a sitcom), space security policy impacts society (Cohen 2014; Netflix Series: Series: Space Force Season One 2020).

The cultural influence of space activities inspires a diverse group of individuals to engage in space exploration but also in space policy and security – but do those leading the charge reflect that diversity?

4.4 Gender Diversity in Space

In 2019 NASA planned the first "all-woman spacewalk" on the International Space Station and then promptly blundered it by not having the correctly-sized space suits available for the astronauts (Popovaite 2020). The walk later occurred, with U.S. astronauts Christina Koch and Jessica Meir, but the initial public mishap is analogous to NASA's long sorted history with gender diversity (Popovaite 2020). Following the space walk in August 2019, the National Space Council convened for the sixth time. The session focused on NASA's new Artemis program and its upcoming mission to land the first woman (and the next man) on the moon by 2024 (US White House 2019). The assembled panel included Saralyn Mark, founder and president of iGIANT and SolaMed Solutions LLC, who addressed how NASA might handle the increased diversity in future crewed flights (Gohd 2019). She counseled,

> It's more than spacesuits not fitting female astronauts. We see the impact from the shoes and clothing we wear, the electronic devices we use, the cars we drive in and even the medications we take. Modifying appearances, or the 'pink it, shrink it' approach for gendered innovation will never work in any environment including space, battlefields, hot zones and in our homes (Gohd 2019).

Many have discussed the importance of studying the differences between male and female physiology in space, as human bodies are built differently with different chemical balances (Bartels 2019). But Mark's comments bring attention to the issue in a broader *outer-space as a theater of war* context, mentioning battlefields to hot zones.

To that point, NASA is far from the worst offender in struggling to figure out how to enhance gender diversity. According to a May 2020 report released by the U.S. Government Accountability Office, in fiscal year 2018 83.5% of the DoD population identified as male (US Government Accountability Office 2020). The report also indicates that the DoD has no real guidance or plans for its efforts to recruit and retain women (US Government Accountability Office 2020). In response to the report, Elizabeth Van Winkle, the Pentagon's executive director of force resiliency, seemed to agree with the recommendations to create (at the very least) some kind of retention plan, saying "While neither the Department of Defense nor the military services set recruitment, accession or retention goals based on race or gender, DoD will continue to study the dynamics of female recruitment, accession and retention." (Mayer 2020). She went on to indicate that increased maternity leave was a recent example of progress (Mayer 2020).

The need to not only attract, but also retain, women plagues the national space security field from the DoD out, but perhaps positive movement is on the horizon for women interested specifically in space security. The newly created Space Force's first two hires were both women, a public move meant to show the service's interest in being gender-neutral from its very beginning (Feldscher 2020). Air Force Maj. Gen. Crosier, the director of planning for U.S. Space Force, said the service is actively seeking ways to incorporate policies that allow for career flexibility and generous family and caregiver leave, and to create a culture of equality and inclusion. When speaking to the Defense Advisory Committee on Women in the Services Crosier said, "We have not yet developed or built what are the core values of the Space Force. I tell you what, diversity and inclusion has got to be part of it from the foundational level." (Hodge Seck 2020).

These comments are a hopeful and significant change in DoD rhetoric on inclusion. The Space Force seems to be paying close attention to the importance of language and terms too. "We are working right now on what we call space professionals. [The new term] won't have 'man' in it. It won't be a term like 'spacemen.' It won't be a term that conjures up gender." Croiser further explained to the committee in 2020 (Hodge Seck 2020).

While these new attitudes and efforts are applauded, there is still a long way to go in finding gender balance in the field – and those of us training the next generation of professionals are setting the stage.

4.5 Efforts in Academia: The University of Nebraska

As discussed in the first portion of this chapter, plenty of women are graduating from non-military undergraduate colleges, and many are going on to reputable graduate programs in national security. Enhanced visibility of women in this sector shifts the local and national perceptions of what it means to serve one's country as a civilian or a military member. Individuals who are passionate about their country come in all genders, colors, backgrounds, and ages – student bodies and the new workforce should reflect that truth.

Universities have an exceptional opportunity to help enact that shift. College is a time when educators have the attention and interest of women; a pivotal period in developing their substantive knowledge, job skills, resiliency, and to nurture a passion that may lead to a lengthy career. Through university collaborations with the DoD and the intelligence community, educators have a unique opportunity to use our time with students – undergraduate, to graduate, to doctoral – to create a pipeline into these career fields for female students.

The intelligence community and the national security sector are also interested in student development and recruitment. A good example is the Intelligence Community Centers for Academic Excellence (IC CAE) program. The IC CAE is a congressionally mandated program established in 2005 with the goal of creating a pool of culturally and ethnically diverse, multi-disciplinary job applicants for the Intelligence Community (IC). The IC CAE programs provides grants to educational institutions to support the design and development of intelligence-related courses (Defense Intelligence Agency n.d.). While the program was initially started within the CIA, The Office of the Director of National Intelligence (ODNI) is responsible for the operation of the program, and the daily oversight is located within the Defense Intelligence Agency (DIA) (Defense Intelligence Agency n.d.).

The University of Nebraska political science department received one of these IC CAE grants and continues to participate in the program (Defense Intelligence Agency n.d.).[4]Undergraduate students may become part of the department's IC Scholars program and even minor in national security, before they even begin thinking of graduate school (University of Nebraska 2020). The IC Scholars program, led by Dr. Tyler White, is particularly engaged in recruiting women. In the 2019–2020 academic year 27 of the 43 students participating in the program were women and the group's executive student leadership team was all female two of the past three years.[5]The group has regular meetings focused on substantive skills (analysis training, bringing in subject matter experts, etc), professional development (practice interviews, resume reviews, etc), and helps to facilitate internships, study abroad opportunities, and eventual job placement. Annually the program takes students to Washington D.C. to visit the recruitment offices of most of the national intelligence agencies and to meet alumni working in the area. Dr. White reflected, "seeing young women feel empowered to take on national security issues facing our nation, and take leaderships roles in our student organizations to encourage their peers, has been a highlight of leading this program."

Beyond the IC Scholars program, Nebraska has a unique and direct relationship to U.S. Strategic Command (USSTRATCOM). As USSTRATCOM's University Affiliated Research Center (UARC), the National Strategic Research Institute (NSRI) at the University of Nebraska provides more than 350 interdisciplinary, experienced researchers with unparalleled ability to investigate and solve multiple defense-related challenges. The UARC essentially partners DoD needs with academic researchers through a dedicated contract vehicle. At the UARC's creation, the Space, Cyber, and Telecommunications Law Program, within the College of Law, was declared one of the core program competencies, showcasing an interest and need in policy and law research.

[4]See schools who have participated.

[5]Data provided by Dr. Tyler White, IC Scholars Program Director.

The Space, Cyber, and Telecommunications Law LL.M. program is also distinctive in its student body.[6]The majority of the student body is made up of Air Force, Navy, Army, and Marine Judge Advocate General (JAG) officers who are attending the program during their legal military career, in active duty status. The relationship, particularly with the Air Force, has been a key part of the program since its beginning. These professional military students add practical perspectives in the classroom setting and as a result are an excellent resource to J.D. students.[7]

In the midst of all this DoD and government engagement, the divisiveness of politics is high and differing views on defense are common throughout college campuses. It's vital the University of Nebraska continue to grow and foster its relationship to USSTRATCOM, as well as broaden program opportunities, and support our military affiliated student-body. It is also vital to the health of the University community that students and faculty on all sides of the political spectrum, and from diverse backgrounds, feel engaged with institutional service to the DoD and in the security and intelligence sector. Because so much of national security law and policy requires detailed knowledge of world-history, international law, and global economics it is a valuable way to engage students in complex, high level discourse; breaking the "me vs. them" mentality and requiring students to identify the motivations and interests of others. These are critical skills for our students in the modern political climate.

One recent project, Women in National Security (WiNS), was developed with just these concerns in mind. Dr. Tyler White and I worked with our Office of Diversity and Inclusion on an internal development grant program. WiNS sought to establish student programing and resources to provide career development resources and field-specific mentorship to women engaged in national security. WiNS focused on three core objectives: visibility of women in the field, skill-building, and mentorship. The one-year project sponsored four student-centric events bringing in prominent women to meet with students, such as a leadership identity workshop lead by *Bossed Up* founder and CEO Emilie Aries to one-on-one resume review sessions with Mary (Beth) Barns, a 32 year career vet from the CIA who received her Career Intelligence Medal at her retirement on January 8, 2019. In all, WiNS supported multiple events within the IC Scholars program and with space law students at the law college, while also inspiring our faculty to think more creatively about how to build professional skills in our students.

---

[6]An LL.M. is a post-doctoral program that follows on to a Juris Doctorate (JD) or general practice law degree in the U.S. Law School Admissions Council. See https://www.lsac.org/discover-law/types-law-programs

[7]A J.D. is a juris doctorate degree, commonly referred to as a law degree. See https://www.lsac.org/discover-law/types-law-programs

4.6 Impacting Our Pedagogy

At the same time Dr. White and I were developing and implementing WiNS, I became interested in what the law college's alumnae employment history might tell me about how our pedagogy and events impact the retention of women in national security positions. I started by sitting down with group of women I really admire—our LL.M. alumnae.[8] I focused on women who have been practicing in the technology, security, or aerospace fields for five or more years. I identified seven women ranging from women I knew very well, including personal friends, to women I only knew professionally. I also reached beyond speaking to women in law and policy and sought input from those in science and engineering more generally. I regularly engage with the U.S. State Department Leadership Program, discussing national space and space policy. The visitors that semester happened to be a diverse group of female scientists. They were traveling the U.S. to explore ways to increase the participation of women and girls in STEM (science, technology, engineering, and math) fields, examine how women contribute to economic growth, and engage with scientists and engineers around the United States who support and mentor women in STEM. I joined this group of scientists from Curacao, Indonesia, Mongolia, Trinidad & Tobago, and the United Kingdom to talk about their experiences and gain their insight into how their college experiences impacted their career choices.

Most of my interviews turned into natural conversations, but I started with a set list of interview questions, such as, "Did your higher-ed experiences prepare you for varied communication styles and other workplace realities?" or "During your education—at any institution—did you receive mentoring on dealing with perceived or actual inequality in the workplace?" In all I have spent over 50 h talking to women about their stories, their love of science and technology, their successes, and their failures. I examined those interviews looking for trends and similarities. From these results I pulled out a few common experiences most of the women faced, and their advice on what skills are necessary in practice.

Every single woman I talked to identified "The Imposter Syndrome" as a hurdle in their careers (Dalla-Camina 2018). The imposter syndrome is generally defined as a professional feeling like an imposter in their position or profession, despite being qualified (Dalla-Camina 2018). I asked women to identify where they thought this insecurity came from and to consider how educators could respond. Many women reported being devastated by one or two specific events they perceived as failures, which in a culture that often expects perfection from women, damaged their confidence for years.[9] This articulated a distinct need for resilience training. When

---

[8]This section adapted from my article in Law Practice Today, an American Bar Association Law Practice Division publication, published 2018 February 13, https://www.lawpracticetoday.org/article/women-aerospace-technology-policy/

[9]For further reading on the culture of female perfectionism, see Debora L. Spar's book, "Wonder Women: Sex, Power, and the Quest for Perfection" published by Sarah Crichton Books, 2013.

emphasis is placed on process and effort, individuals begin to value their skills over their output (Educational Leadership 2013). All humans will fail at something, but the more focus placed on the skills that have been developed (over the successful or failed outcome) the more resilient people become, and the more likely they are to try again (Educational Leadership 2013). We determined it is vital to give students multiple, low-risk opportunities to fail and then assess that failure, rather than try to hide it or shame them for it. Another way leaders and educators can do this is to share our own failures and inadequacies. This does not cut down our authority or prestige; rather, it humanizes us and shows others that failure is not final.

As for skill building, a few ideas were mentioned again and again by the women I interviewed. Many of the women talked how critical is was to be able to accept negative feedback without internalizing it as personal attacks – ultimately going back to building resilience. One way to implement this is to stop delivering critiques, grades, and feedback in writing only. Written feedback is standard practice at most law schools and in many firms and agencies, but educators should consider taking time to talk through feedback with students, encouraging them to practice how they respond.

Other skills-building recommendations included the power of role-play in the classroom (such as interviews or intelligence briefings), teaching women tactics for self-promotion (such as practicing how they talk about their interests and networking), and the significant importance of public speaking.[10]Public speaking, from large classes to briefing a student group on a topic, is a simple way to help women develop confidence. Practicing the art of articulating complex information in front of others, often on the spot, is a vital skill educators can help students hone.

Most of the women also reported struggling to advocate for themselves early on in their careers, which is also where they saw many of their peers stall out. This lack of self-advocacy was as simple as feeling embarrassed at networking mixers to failing to negotiate their employment offers. In fact, more than half of the women I spoke with had never asked for a raise in their entire career. In response, the team at Nebraska has created multiple roleplay scenarios and scripts for students to practice these conversations. Also, many professional associations offer roleplay exercises specific to salary; in fact, the American Bar Association has a robust library of exercises and an entire workbook dedicated to the topic.[11]

I also learned that networking is a dirty word for many professionals. Countless students and recent alums tell me that networking is awful and feels uncomfortable. Most seasoned professionals can empathize with that sentiment but are quick to point out the importance of this skill. I challenge our students to adopt a positive view of networking. If they approach networking from the perspective of how they

[10]A great place to turn for some roleplay exercises and discussion prompts on building grit and resiliency strategies is the American Bar Association Grit Project. The Grit Project aims to educate women lawyers about the science behind grit and growth mindset. The project's toolkit provides discussion scenarios such as how to handle speaking up in class, the job search, and a bad grade on a midterm paper.

[11]See https://www.americanbar.org/groups/diversity/women/initiatives_awards/grit-project/

may be able to help others, rather than what can others give them, it can be become a very rewarding experience. We suggest nervous networkers do two things. First, prepare a few things to say about yourself and your career interests. Second, and much more importantly, practice engaged listening and come prepared to ask people questions about their careers. Go into networking events looking for ways to connect other people and build relationships by helping others reach their goals. This sort of relationship building is meaningful, reflects positively on everyone involved, and often pays off in unexpected ways.

Students sometimes tell us, "These are accomplished attorneys/analysts/professors, how could I possibly be helpful to them?" I remind them that they may know their next big client, or give them advice on their next vacation, or refer them to their kid's new favorite summer camp. Everyone has the capacity and connections to help another person. Also, thankfully, this is another skill that can be practiced. Our Career Development Office hosts networking practice sessions – which comes with the added benefit of students getting to know each other better.

Finally, every interview I did also eventually landed on mentorship and sponsorship – needing one and being one. The differences between mentors and sponsors can seem minimal, but the two are actually quite distinct:

> A mentor advises the mentee; sponsors advocate for their protégés. A mentor can be anyone in a position of experience, while a sponsor is a senior-level executive. Mentors help build a career vision. Sponsors drive their protégé's career vision. Mentors suggest ways to expand the mentee's network. Sponsors give protégés their active network connections and make new connections for them. Sponsors are personally vested in the upward movement and professional development of their protégé. They champion their protégé's visibility. (McDaid 2019).

A Catalyst study found that sponsorship has become more influential than mentoring in helping women succeed in business (McDaid 2019). A sponsor actively advocates for their protégé when they are not in the room and when high level decisions are being made. It is a high stakes relationship and difficult to form. Experts suggest that above and beyond showing discipline and being a problem solver, women must be able to clearly articulate and share their goals (Giang 2019). This calls directly back to the necessary skills associated with self-promotion: resiliency, networking, and public speaking. We must teach women to promote themselves and practice it.

As for mentors, almost none of the women I interviewed cared about what gender their career mentors were. What they sought was someone with specialty-specific guidance as to the most important events, certifications, and publications. In discussing this with my one of my mentors, Nebraska Law Assistant Dean Molly Brummond, she shared some great advice I pass on to our students: bypassing the "IRL" (in real life) relationship with high profile role models. In addition to pursuing in-person relationships, we encourage students to identify professionals in their field that they respect and want to emulate and to follow them on social media. Reading what they read, looking at who they interact with, and taking note of what conferences they're tweeting from. Students will begin to get the lay of the land on their own. We encourage them to take it a step further and to reach out to them informally—a quick note thanking them for posting about a book they found helpful or an article they enjoyed.

Similarly to our pedagogical and career development assessment interviews, the #NatSecGirlSquad survey asked women what they would like to see change in the sector to support women (#NatSecGirlSquadxGuidehouse 2020). The results reflected many of the same experiences and advice but one additional piece of feedback focused on the job application process. "In my experience, much of the application process is unknown. It would be helpful to talk to people who have been through the process or have access to resources that better prepare me for the process so that I know generally what to expect and how to best prepare myself" one responded shared (#NatSecGirlSquadxGuidehouse 2020).

This is where alumni mentorship becomes critical for young women. Giving them the "lay of the land" not just for day to day office life, but for what are often lengthy and complex hiring processes. For example, we created a step by step best practices guide for using the website USAjobs.gov, the federal governments hiring page in addition to referencing many guides to the site (USAJOBS 2020). Last year we had previous NASA interns host a panel discussing their application process and managing the portal. We do similar partnering and coaching for students or recent alumni going through the security clearance process. These are simple connections that can deeply impact access to the field.

Turning to experiential education and internships, one #NatSecGirlSquad survey taker said,

> The national security field relies heavily on personal connections for job opportunities, and especially on unpaid internships for entry-level positions. This systematically harms people who can't afford to take those positions or who attended schools that are not directly plugged into those personal networks. I think this is one of the biggest reasons for the lack of diversity in our field. (#NatSecGirlSquadxGuidehouse 2020).

Reminding educators that not all students arrive in our classrooms on the same footing, and that internships are a fundamental educational tool. In fact, 76% percent of employers cited relevant work experience as the primary factor in hiring decisions (Edwards et al. 2010). In 2010 researchers estimated the cost of the three month internship in Washington D.C., excluding travel at $4050 (Edwards et al. 2010). Students interested in nonprofit or government positions have a particularly difficult time finding paid internship opportunities. Unlike for-profit firms, nonprofit organizations and government agencies are not legally obligated to pay their interns and many report that the simply can't (Mayo and Shethji 2010). Unfortunately, students who may be supporting themselves, or their family, often are not able to afford a summer or semester working for free.

Many Universities have created programs within their colleges to help support these students. For example, the University of Nebraska operates the Nebraska Public Interest Law Fund. The fund, supported by donations and an annual student auction, help cover living expenses for students working at unpaid positions in the public interest. However, many programs lack the full resources to accommodate the student need (Nebraska College of Law 2020). Groups like Pay Our Interns have come in to attempt to fill those gaps. Pay Our Interns is a bipartisan, non-profit organization that advocates for an increase in the amount of paid internships within government, for-profit, and nonprofit sectors (Pay Our Interns 2020). Additionally the group serves as a research platform, advocacy guide, and operates a paid internship database.

The complexity of income disparity in job readiness programs and internships is a developing and difficult issue educators face, but it can not be ignored as we think creatively about expanding diversity in national security.

It, along with the other pedagogical strategies discussed here, are just one piece of a very complicated recipe in supporting women interested in national security space careers.

4.7 Conclusion

This look at nation-wide data and an overview of individual personal interviews with professional women paint a clear picture: we're a long way off from gender equality in the national security and space sectors. However, the United States is poised to change that now more than ever before. From the "business case for diversity" research trend, women-focused national security professional organizations, increased female student interest, and educational efforts there is a growing wave of professional focused on gender equality.

Considering the success of the IC Scholars program, our UARC, and our student body, the University of Nebraska community is uniquely posed to make significant inroads into supporting diversified interest in national security jobs among female students. Furthermore, educators at Nebraska (and beyond) have a responsibility to do more than just deliver the subject matter to our students. Universities must prepare competent female professionals with the resilience and skills that equip them for success.

It is my sincere belief that someday, when my career nears retirement, I will have to stand in many very long bathroom lines.

References

#NatSecGirlSquadxGuidehouse. (2020). *What it looks like vs. What it is: Building competent diversity in national security (survey)*. https://www.natsecgirlsquad.com/survey

Air Force Space Command. (n.d.). *Air Force Space Command History*. https://www.afspc.af.mil/About-Us/AFSPC-History/

Bartels, M. (2019, April 18). Second NASA astronaut to spend a year in space – For science. Space.com. https://www.space.com/christina-koch-year-in-space-science.html

Byron and Post. (2016). Women on boards of directors and corporate social performance: A meta-analysis. *Corporate Governance: An International Review, 24*(4), 428–442.

Catalyst Staff Advisors. (2004). *Report: The bottom line: Connecting corporate performance and gender diversity.* https://www.catalyst.org/research/the-bottom-line-connecting-corporate-performance-and-gender-diversity. Accessed 20 May 2020.

CNAS. (2016, June 20). *CNAS survey administered at the 2016 Annual Conference.*

CNN Editors. (2013). By the numbers, women in the U.S. military. *CNN.* http://www.cnn.com/2013/01/24/us/military-women-glance. Accessed 27 July 2020.

Cohen, N. (2014, September 13). We're all nerds now. *The New York Times.* https://www.nytimes.com/2014/09/14/sunday-review/were-all-nerds-now.html

Dalla-Camina, M. (2018, September 3). The reality of the imposter syndrome. *Psychology Today.* https://www.psychologytoday.com/us/blog/real-women/201809/the-reality-imposter-syndrome

Defense Intelligence Agency. (n.d.). *Frequently asked questions.* https://www.dia.mil/Training/IC-Centers-for-Academic-Excellence/Frequently-Asked-Questions/

Duffin. (2020). Number of bachelor's degrees earned in the United States from 1949/50 to 2028/29, by gender. *Statista.* https://www.statista.com/statistics/185157/number-of-bachelor-degrees-by-gender-since-1950/. Accessed 27 July 2020.

Educational Leadership. (2013, September). Resilience and learning. *Educational Leadership, 71*(1).

Edwards, K. A., Hertel-Fernandez, A., Cauthen, N. K., & Economic Policy Institute. (2010). *Paving the way through paid internships: A proposal to expand educational and economic opportunities for low-income college students.* Dēmos: Economic Policy Institute.

Erwin, S. (2019, October 23). Five things to know about U.S. Space Command. Space-News.com. https://spacenews.com/five-things-to-know-about-u-s-space-command/. Accessed 20 May 2020.

Fairchild, C. (2014, June 3). Number of fortune 500 women CEOs reaches historic high. Fortune.com. http://fortune.com/2014/06/03/number-of-fortune-500-women-ceos-reaches-historic-high. Accessed 27 July 2020.

Feldscher, J. (2020, March 5). Space Force leader's quest: A gender-neutral force. Politico.com. https://www.politico.com/news/2020/03/05/space-force-gender-neutral-122556. Accessed 27 May 2020.

Giang, V. (2019, February 26). 7 successful women explain how they got the sponsor that changed their careers. *Fast Company.* https://www.fastcompany.com/90307513/7-successful-women-explain-how-they-got-the-sponsor-that-changed-their-careers

Gohd, C. (2019 August 21). If NASA wanted to the first woman on the moon her spacesuit better fit. Space.com. https://www.space.com/nasa-gender-specific-mike-pence-nsc.html. Accessed 27 May 2020.

Goodwin, Chemerinsky, & Erwin. (2019). The transgender military ban: Preservation of discrimination through transformation. *Northwestern University Law Review, 114*(3) UC Irvine School of Law Research Paper No. 2019-37.

Hodge Seck, H. (2020, March 3). US Space Force may become the most female-friendly military services. Military.com. https://www.military.com/daily-news/2020/03/03/us-space-force-may-become-most-female-friendly-military-service.html. Accessed 27 May 2020.

Kidder, Schafer, Carter, & Swick. (2017). From college to cabinet women in national security. *Center for a New American Security Report.* https://www.cnas.org/publications/reports/from-college-to-cabinet. Accessed 27 July 2020.

Lupton, D. (1998). *On space warfare: A space power doctrine.* Maxwell Air Force Base, Alabama: Airpower Research Institute. Air University Press.

Lublin. (2016). Lublin article: https://www.wsj.com/articles/companies-try-a-new-strategy-empathy-1466501403

Mayer, M. (2020, May 21). The military needs a real plan to recruit and retain women, report find. *Military Times.* https://www.militarytimes.com/news/your-military/2020/05/20/the-military-needs-a-real-plan-to-recruit-and-retain-women-report-finds/. Accessed 27 May 2020.

Mayo, L., & Shethji, P. (2010). Reducing Internship Inequity. *Association of American Colleges and Universities, Diversity and Democracy, 13*(3).

McDaid, E. (2019, August 14). Mentor vs. Sponsor. *The Leader's Edge*. https://www.leadersedge.com/brokerage-ops/mentor-vs-sponsor
McMahon. (1983). *Memorandum from CIA Deputy Director*. http://www.foia.cia.gov/sites/default/files/document_conversions/1820853/1983-12-15.pdf. Accessed 27 July 2020.
NASA. (2018). *What is NASA?* https://www.nasa.gov/audience/forstudents/5-8/features/nasa-knows/what-is-nasa-58.html. Accessed 20 May 2020.
Nebraska College of Law. (2020). *Nebraska Public Interest Law Fund*. https://law.unl.edu/NPILF/
Netflix Series: Space Force Season One. (2020, May 29). https://www.netflix.com/title/81021929
Noland, Moran, & Kotschwar. (2016). *Is gender diversity profitable? Evidence from a global survey*. Washington, DC: Peterson Institute for International Economics.
Offermann, & Foley. (2020). *Is there a female leadership advantage?* Oxford Research Encyclopedia, Business and Management.
Office of Space Commerce. (n.d.). *National Space Council*. https://www.space.commerce.gov/category/national-space-council/
Oleson, Silsby, & Skelly. (2012). *The space review. NASA is essential for national security*. https://www.thespacereview.com/article/2210/1. Accessed 27 July 2020.
Pay Our Interns. (2020). https://payourinterns.org/
Popovaite, I. (2020, January 14). What the 2010s taught us about women in space. SpaceNews.com. https://spacenews.com/what-the-2010s-taught-us-about-women-in-space/
Smith, M. (2018, March 14). Top Air Force officials Punt on Trump's Space Force. SpacePolicyOnline.com. https://spacepolicyonline.com/news/top-air-force-officials-punt-on-trumps-space-force/
Smith, M. (n.d.). Military/national security space activities. SpacePolicyOnline.com. https://spacepolicyonline.com/topics/militarynational-security-space-activities/. Accessed 3 Aug 2020.
United States Space Command. (2020, February 26). *Fact sheet*. https://www.spacecom.mil/About/Fact-Sheets-Editor/Article/1948216/united-states-space-command-fact-sheet/
University of Nebraska. (2020). *National security studies*. https://nationalsecurity.unl.edu/
US Government Accountability Office. (2020, May 19). *Guidance and plans needed for recruitment and retention efforts*, GAO-20-61.
US White House. (2019, August 20). *Remarks by Vice President Pence at the sixth meeting of the National Space Council*. https://www.whitehouse.gov/briefings-statements/remarks-vice-president-pence-sixth-meeting-national-space-council/
USAJOBS. (2020). https://www.usajobs.gov/
Weeden. (2019). *U.S. national security space policy and strategy: From sanctuary to space force*. Presented at the JMOD Symposium on Space Security, Tokyo, Japan, February 6, 2019.
Windrem, R. (2013, November 14). Hunting Osama bin Laden was women's work.NBCNews.com*Investigations*. https://www.nbcnews.com/news/world/hunting-osama-bin-laden-was-womens-work-flna2D11594091
USSPACECOM https://www.spacecom.mil

Chapter 5
Rules of the Domain: The Impact of Law on the Evolution of Space Activities and Technologies

Susan Trepczynski

Abstract Understanding space law, including the development and evolution of international space law, as well as the interplay between international legal obligations and domestic law, is key to understanding the fundamental bases on which space activities are prioritized, integrated, and utilized in a given State. The domestic legal regime in place in a State may have a significant impact on the development of space activities and emerging space-related technologies, which can be positive (enabling), negative (constraining), or neutral. Whether a legal system places intentional constraints on particular activities, does not offer adequate legal certainty for those looking to engage in or enable innovative activities in the space domain, provides minimal regulation, or affirmatively encourages development of space technologies and activities, for those developing emerging technologies, seeking to engage in innovative space activities, or advising those who do, it is critical to understand how the applicable legal regime will impact the activity or technology – from conception to execution.

5.1 Introduction

When told that I specialize in space law, a common reaction is "there's law in space?" Granted, as space-based technologies become more engrained in our everyday lives, space activities become more visible, and the significance of space as a domain is discussed more regularly in the media, the frequency of this reaction has gone down – but not as significantly as one would imagine. Why are so many people surprised to find there is law in space? Perhaps it is because there is not yet a large human presence in space, and many people relate law to human activities. Perhaps it

The views expressed herein are solely those of the author and do not represent the views of the DoD or any of its Components.

S. Trepczynski (✉)
United States Air Force, Washington, DC, USA

M. de Zwart, S. Henderson (eds.), *Commercial and Military Uses of Outer Space*, Issues in Space, https://doi.org/10.1007/978-981-15-8924-9_5

is because space seems so distant to many people, even though it is, in reality, only about 100 km away (less than an hour drive on a highway). Perhaps it is because the utility of, and benefits derived from, space are not always widely recognized, though the increasing visibility of commercial space is certainly changing that. Perhaps it is because space still carries a sort of mystique for many, invoking images from science fiction stories. Whatever the reason, the fact the mere existence of law in space is surprising to many, underscores the importance of increasing understanding of the important role law plays in space.

Generally speaking, there are two categories of law that govern and impact activities in space – international law and domestic law. International law contains a broad set of principles applicable to State activities in space, while domestic law sets the rules that constrain or enable, sometimes in intricate regulatory detail, space activities in a particular State. Whereas international law sets the outer boundaries of State activities, domestic law provides insight into how a State views space across a multitude of perspectives, to include political, military/national security, economic, and commercial. Understanding space law, including the interplay between international legal obligations and domestic law, is key to understanding the fundamental bases on which space activities are prioritized, integrated, and utilized in a given State. Space law represents the rules of the space domain, a domain complicated by the fact the rules are not uniform. What is legally possible for those falling under the jurisdiction of one State may not be legally possible for those falling under the jurisdiction of another. However, in all cases, whether looking at innovative technologies or emerging activities, the applicable law sets the boundaries for what is possible and, in some cases, evidences those aspects that must be changed (or evolved) to ensure the rules of the domain keep pace with its growing importance and expanding utility.

5.2 Development and Role of International Space Law

On 4 October 1957, the Soviet Union (USSR) launched *Sputnik I*, the first artificial satellite, into low Earth orbit, marking the first human activity in space and the beginning of the space age. At the time, there was no international law specifically applicable to space and certainly no international law governing human activities in space. While plans to orbit an artificial satellite had clearly been in the works in both the USSR and the United States (U.S.) for a number of years prior to the *Sputnik I* launch, the lack of a legal framework was not surprising.[1]There was obviously little reason for States to devote resources to developing law to govern a domain that was devoid of human activity. From a more pragmatic perspective, developing law in anticipation of activities that were secretive, intimately connected to a greater

[1]In 1955, both the U.S. and the USSR publicly announced an intent to launch satellites into orbit in connection with the 1957–58 International Geophysical Year (IGY).

geopolitical conflict, and linked to national security was also untenable. However, the launch of *Sputnik I*, closely followed by *Sputnik II*, and the January 1958 launch of *Explorer I* by the U.S., changed everything. Space capabilities were revealed to the world, space suddenly became a domain of potential use to humans, and, perhaps more significantly at the time, space took on a new importance as another aspect of competition between the two Cold War superpowers. It was against this backdrop that the United Nations (UN) passed Resolution 1348 (XIII), establishing the Committee on the Peaceful Uses of Outer Space (COPOUS), which was tasked with, among other things, examining the types of legal problems that might arise in relation to space exploration.[2]

While COPOUS was considering the legal implications of space exploration, the fact States were simultaneously engaging in various space activities carried legal significance. For example, although the question of where airspace ends and outer space begins remains unsettled in international law to this day, the launch of *Sputnik* established that States perceived outer space and air space as separate domains, governed by distinct legal regimes. This was evidenced by the lack of State protest when *Sputnik* (and subsequent satellites) passed over the sovereign territory of other States. Under international air law, States exercise sovereignty over the airspace above their territory, meaning an aircraft following the rough path of Sputnik (albeit necessarily at a much lower altitude) would have been in violation of the sovereign airspace of many States. However, despite the lack of a fixed boundary between airspace and outer space, no State protested the 'overflight' of Sputnik. Given this lack of protest, there seemed to be an implicit acceptance by States, right from the onset of human space activities, that the rules applicable to the air and space domains were different. This acquiescence has been described as creating customary international law with regard to the freedom of use of outer space (McDougal 1963).[3]

With State practice establishing a *de facto* legal distinction between airspace and outer space, the USSR and the U.S. continued their space activities (to include first manned launches in April and May 1961, respectively), and COPUOS pursued the question of the legal implications of those activities. In 1963, this resulted in the UN General Assembly *Declaration of Legal Principles Governing the Activities of States in the Exploration and Uses of Outer Space*, representing a first attempt by States to articulate general guiding principles for space activities. While the Declaration was not itself legally binding, the principles influenced, and were largely incorporated into, the subsequent binding treaties making up the body of international space law as we know it today.

The four primary treaties that comprise the international law governing space and space activities are the Outer Space Treaty; Rescue and Return Agreement; Liability

[2]Resolution 1348 (XIII) established COPUOS as an ad hoc committee, but its existence as a permanent body was formalized a year later by UNGA RES 1472 (XIV), 12 Dec 1959.

[3]Specifically noting freedom of access "would appear today to be already, without formal convention, an accepted principle of international law".

Convention; and Registration Convention.[4] These treaties, the last of which entered into force in September 1976, are very much reflective of the time and political climate in which they were negotiated and concluded; none of them have been amended since. They are a product of the 'space race' that developed between the USSR and U.S., a contest that recognized the national security and foreign policy significance of space exploration and technology, serving to accelerate the pace of both. The treaty negotiations were a forum in which the superpowers could pursue their respective agendas, and the resulting law had to be agreeable to both.

Given the backdrop of the Cold War, it is not surprising the treaties do not place significant restrictions on the space activities of States, and make almost no specific mention of military activities. The treaties, none of which is particularly lengthy or detailed, outline broad principles and general areas in which international consensus, and, more critically, consensus between the superpowers could be reached. The Outer Space Treaty itself is a sort of umbrella agreement, establishing foundational legal concepts, some of which were subsequently expanded upon in the Rescue and Return Agreement, Liability Convention, and Registration Convention. This permissive international legal regime, with many provisions that are open to interpretation, was ultimately acceptable to both the USSR and the U.S., allowing each to continue to pursue multifaceted domestic space programs.

The permissive nature of international space law is one of its fundamental characteristics. The primary purpose of international space law is to set the outer boundaries of space activities, defining the broad rules of the domain for States that choose to undertake space activities, and guaranteeing all States the right to engage in the use and exploration of space. International space law does not require States to have any particular sort of space program or any space program at all. It does not prescribe specific activities or the entities within a State that may participate in them. However, it does set a baseline for those activities (at least for those States' Parties to the international treaties, and perhaps more broadly to the extent some provisions now reflect customary international law),[5] and understanding that baseline is critical to understanding and advising on space activities.

---

[4] A fifth treaty, the Moon Agreement, does not have wide adherence. As of January 2019, there were 18 States Parties to the Moon Agreement, none of which were States with orbital launch capabilities. A compilation of all these treaties, as well as other related legal documents, is available from the United Nations Office of Outer Space Affairs (UNOOSA), 2017, *International Space Law: United Nations Instruments*.

[5] States are only obliged to comply with treaties to which they are party unless the treaty (or particular provisions of it) reflects customary international law. Customary international law is generally defined as requiring two elements – general State practice and acceptance as a legal obligation (i.e., the practice is undertaken because States believe they are legally obligated to act in accordance with the practice) (International Law Commission, 2018, *Draft Conclusions on Identification of Customary International Law*). While views on the customary status of the space treaties, or specific provisions of them, currently vary from State to State, the overall impact of the treaties on the activities of States can also be assessed by considering the fact the vast majority of States with space capabilities (and especially those with launch capabilities) are party to the OST, Rescue and Return Agreement, Liability Convention, and Registration Convention.

Two of the foundational principles of international space law that contribute to its permissive character are freedom and non-appropriation. Article I of the OST establishes all States are free to use and explore space, to include free access to all areas of celestial bodies, as long as such use and exploration is in accordance with international law. This freedom is reinforced by the non-appropriation principle found in Article II of the OST, which prohibits States from making sovereign claims, whether by use or occupation, to outer space or any celestial body; neither outer space as a whole, nor any part of it, is subject to the jurisdiction of any State. The OST also mandates the application of international law, including the UN Charter, to outer space.[6]Taken together, these provisions make clear that States retain a significant amount of freedom in determining the specific types of space activities they will engage in, so long as those activities are lawful under international law.

The OST also establishes several notable international legal obligations, some of which are expanded upon in the subsequent treaties (i.e., the Rescue and Return Agreement, Liability Convention, and Registration Convention).[7]Article IV limits the placement of nuclear weapons or weapons of mass destruction in space, and limits the military activities that can take place on celestial bodies.[8]Article V provides protections for astronauts, directing States Parties to treat them as "envoys of mankind" and obligating States to provide them "all possible assistance."[9]Article IX requires States to "conduct all their activities in outer space ... with due regard to the corresponding interests of other States Parties."[10]

The OST also contains several provisions which, given their purpose of addressing rights and obligations related to space objects and activities, have importance in both international law and, with respect to how States implement international obligations, in domestic law. The first these provisions is the concept of a "launching State," which is legally significant because the OST assigns international liability to launching States[11](complicated by the fact any given launch may result in more

[6]While it may seem counterintuitive for an international treaty to say that international law applies, the significance of this provision is it makes clear that, to the extent the OST does not address every legal issue that could arise in relation to space, States are to look to international law to fill that void.

[7]Many of the concepts contained in the OST were first introduced in the non-binding *Declaration of Legal Principles Governing the Activities of States in the Exploration and Uses of Outer Space*. Thus, while these concepts were not 'created' by the OST, they are established as binding legal obligations by the OST (acknowledging some argue the 1963 Declaration of Legal Principles may reflect customary international law).

[8]Article IV is the only provision in the space treaties that directly addresses military activities in space.

[9]The content of Article V is expanded upon in the Rescue and Return Agreement, though the Rescue and Return Agreement is, of course, only binding on those States which are a Party to it (or to the extent any of its provisions reflect customary international law).

[10]There is little in the way of State practice that directly lends context to the meaning of "due regard" as it is used in Article IX.

[11]The Liability Convention expands upon the liability provisions found in OST Article VII, but the concept of liability attaching to the launching State remains the same. The scope of liability

than one State meeting the definitional criteria of "launching State").[12]Jurisdiction and control (a concept central to legal rights related to an object) over a space object, however, rests with the State of registry, which is not legally required (by the OST) to also be a launching State.[13]Finally, under Article VI of the OST, States are responsible for "national activities in outer space," regardless of whether such activities are carried out by governmental or non-governmental entities. Article VI further requires the "appropriate State" to authorize and continually supervise the activities of non-governmental entities in outer space, therefore establishing an international legal obligation for States to exercise a certain amount of oversight of all national space activities. The international liability incurred by launching States, along with the affirmative requirement to authorize and supervise, and the legal implications of international responsibility falling upon the State for all national activities, including those of non-governmental entities, provide significant incentive for a State Party to implement a robust domestic legal and regulatory regime that is aligned to the State's tolerance for international risk and attuned to the State's strategic goals and priorities with respect to space.

is not necessarily the same for States that are only Party to the OST and those that are also (or only) Party to the Liability Convention. For example, under the OST launching States are "internationally liable for damage" caused by their space object "on the Earth, in air space or in outer space." The OST does not specify the legal standard for the liability. However, the Liability Convention differentiates between absolute and fault-based liability, depending on the geographic location where the damage occurred. Specifically, States are "absolutely liable" for damage caused "on the surface of the Earth or to an aircraft in flight," but for damage caused elsewhere (i.e., in outer space, including on a celestial body), States are liable "if the damage is due to its fault of the fault of persons for whom it is responsible." These provisions are found in the Liability Convention Articles II and III.

[12]OST, Article VII. Article VII defines a launching State as the State or States: (1) that launch an object into space; (2) that procure the launching of an object into space; (3) from whose territory an object is launched into space; or (4) from whose facility an object is launched into space. Based on these criteria, it is not unusual to have multiple launching States for a particular launch. For example, given the limited number of States with indigenous launch capabilities, it will often be the case that one State will have to procure a launch from another.

[13]Provision found in OST, Article VIII. However, under the Registration Convention (Article I(c)), "State of registry" is defined as "a launching State on whose registry a space object is carried", thus requiring the State of registry also be a launching State. This distinction between the use of terminology in the two treaties serves as an important reminder not to assume obligations are consistent across the treaties. For example, a State that is only Party to the OST could potentially register and exercise jurisdiction and control over an object for which it was not a launching State, while a State Party to the Registration Convention (or the OST and Registration Convention) could not do so unless it was also a launching State.

5.3 Role of Domestic Space Law

As a subset of public international law, international space law has great importance to States because it establishes their international legal obligations, setting the outer boundaries of what is legally permissible with respect to activities in space. International space law does not, however, obligate non-State entities (e.g., individuals, corporations, private organizations). Consequently, for those entities seeking to engage in space-related activities or operations, the generally permissive nature of international space law is only relevant to the extent the State they choose to operate in (or are nationals of) has not placed any additional limitations on such activities.[14] The space law of most primacy and relevance to non-State actors will be the domestic space law applicable in the State(s) in which, or under the authorities of which, a non-State actor operates.

For States, domestic law serves the dual purpose of implementing both international obligations and domestic policies. Because international space law is binding only upon the State itself, it is up to each State to ensure activities under its jurisdiction, or carried out by its nationals, are consistent with its international space law obligations, which is primarily accomplished by authorizing and/or constraining the activities of those entities through domestic law. Therefore, the wide range of what is possible under international law can narrow considerably under domestic law, as States determine the parameters of the legal and regulatory regimes most conducive to their domestic and international priorities with respect to space and space-related industries and activities. Some States may have national policies that maximize the freedoms provided by international space law, while others may significantly curtail the range of permissible activities, especially activities of non-State entities. As a result, for those entities involved in private or commercial space activities, understanding the driving forces behind a State's domestic laws and policies is critical as they navigate the realm of the possible in the State(s) in which they are, or are considering, operating.

On one end of the domestic regulation continuum are those States which have no laws whatsoever directed specifically toward space activities.[15] On the other end are those States with advanced legal and regulatory regimes, encompassing a broad range of space-related industries and activities. Many States occupy the middle ground, with basic laws sufficient to implement their international obligations and

[14] Even in cases where a State has not implemented any (or very little) domestic legislation, compliance with international space law (to the extent any of it represents customary international law and/or the State is Party to any of the relevant treaties) continues to remain a responsibility of the State. Individuals and/or private entities have no legal obligations stemming directly from international space law itself.

[15] For those States that are not party to any of the international space law treaties, there is obviously no need to have any sort of domestic law to implement international obligations (other than those obligations that arise out of customary international law). However, not all States that are party to one or more of the treaties have specific domestic space law or policy. UNOOSA has published a collection of national space law (UNOOSA, 2020, *National Space Law Collection*).

some specific law directed toward activities or industries of particular interest to the State. Activities that are often well-regulated at the State level include launch and communications (in this context, specifically space-based or enabled communications, such as communications satellites), with the more robust regulations in these areas being attributable both to the relatively well-developed nature of these activities and the associated international legal obligations.

The first object launched into space was a communications satellite. Thus, both launch and the use of space for communications have been around since the dawn of the space age, with all subsequent development of space law (at least implicitly) needing to account for these basic space activities. Because international law does address these longstanding uses, it is not surprising to see, in many cases, corresponding well-developed domestic legislation. States with established launch capabilities and, in some cases, fairly robust launch industries, have long had an incentive to develop domestic legal regimes that account for and address the international legal implications of qualifying as a launching State (e.g., international liability and registration obligations), while fostering domestic launch capabilities to the extent desired by the State.

Communications regulations, on the other hand, are fairly consistently implemented on the domestic level by almost every State due to obligations stemming from membership in the International Telecommunication Union (ITU), which has near universal membership.[16]In addition to abiding by the ITU Constitution and Convention, the ITU requires member States to adhere to the Radio Regulations, some provisions of which directly regulate space-based communications (e.g., radiofrequency spectrum use for space and earth stations, as well as orbital slots). Consequently, consistent with the purpose and intent of the ITU, domestic regulations relating to spectrum use and management are directly tied to the international regulations set by the ITU.[17]This provides an internationally coordinated approach to the use of spectrum, to include its use in and through space, as well as the use of orbital slots, which directly impacts the domestic regulations of all member States, driving a highly detailed and prescribed approach to communications.[18]

Launch and communications are, of course, not the only categories of space-related law found in domestic legal systems, but they serve as examples of cases where certain activities may be highly regulated at the domestic level due, at least in part, to overarching international legal obligations. However, it is equally possible for a State to place significant regulations on space-related industries

[16]The ITU presently has 193 Member States (ITU, 2020, *List of Member States*).

[17]For example, the United States telecommunication regulations state the ITU Radio Regulations "have been incorporated to the extent practicable." (7 CFR $2.100, *International Regulations in Force*) Similarly, the *Canadian Table of Frequency Allocations*, states "the Canadian Table differs, where necessary, from the ITU Table." (Innovation, Science and Economic Development Canada, *Canadian Table of Frequency Allocations (2018 Edition)*).

[18]This is not to imply that all States Parties to the ITU have domestic regulations that are identical to the ITU Radio Regulations, as the ITU Constitution itself (in the preamble) recognizes "the sovereign right of each State to regulate its telecommunications."

or activities that have no relation to international obligations (other, of course, than the ever-present obligation for a State Party to the OST to authorize and continually supervise the activities of its nationals). States have many reasons to permit, regulate, and/or prohibit activities taking place within, or by persons subject to, their legal jurisdiction. In some cases, restrictions or prohibitions may be based on internal policy decisions, perhaps driven by purely political considerations, such as prohibiting or restricting commercial involvement in activities or technologies that are perceived as having significant military utility. On the other hand, some States may have policies encouraging those very activities, seeing an advantage in commercial development of potential dual use technologies. Some States may heavily regulate or restrict activities for predominantly economic reasons, based on concerns related to international liability, as well as the domestic costs involved in establishing a sufficient regulatory or oversight regime. Other States may seek to avoid the potential impact of the international liability rules by constructing a system that passes all the risk on to those undertaking the activity, which in turn will impact the cost analysis and risk assessment for any entity considering involvement in space activities. Regardless of the reasoning behind a particular State's domestic law and regulation, it is clear the nature and scope of such laws and regulations can vary widely from State to State, creating a potentially interesting disparity in the space capabilities and activities of States, especially from a commercial perspective.

Whatever the policy considerations behind a particular domestic space law regime, the existence of a well-developed regime (even a highly restrictive one) carries benefits. While a highly developed domestic legal regime will likely place numerous requirements (e.g., license, safety, environmental, and similar) on those undertaking space activities, the structure provided by that legal system can reduce risk associated with the underlying activities, while providing some certainty that activities undertaken in compliance with the law will receive legal protections in the State. Given the cost of entry to the space domain, the importance of legal certainty to concept or technology development cannot be understated. If there is any question as to whether a given activity will be permitted by and/or receive specific legal protections by the State in question, it will likely directly impact the viability of any proposal, especially those that depend on outside investment capital to move forward.

In addition to implementing domestic law for purely domestic purposes, States may also use domestic law to influence perception and interpretation of international law, especially in areas where international law is potentially unclear, such as activities or technologies that did not exist (and therefore could not have reasonably been contemplated by the drafters) at the time the international law was developed. While there are certain well-established uses of space and space technology, those States (and the nationals of those States) that are pushing the realm of what is technologically possible and demonstrating new uses of space, may also be leading the development of international law. Because State practice is a significant factor in interpreting international law (especially in those areas where existing law is silent on a particular activity), States may have much to gain by supporting, or potentially hindering, the development of certain emerging technologies or innovative space activities.

5.4 Law and Technology

Space technologies are steadily becoming more ubiquitous and commercially viable (i.e., potentially profitable), making the utility of (and access to) such technologies increasingly important to various interests of, and within, States. Although space activities were born out of military and geopolitical competition and space remains a domain of great importance for the political and military interests of States), it has also become a domain of increasing commercial utility and economic importance. Commercial ventures are pushing development in sectors and technologies that were once almost exclusively within the realm of governments. Because States are no longer controlling the development of space-related technologies, they are being forced to address the legal issues raised by such technologies, often on timelines they did not anticipate, as they attempt to balance national priorities and interests with commercial development and interests of the private sector.

As the scope of activities and the range of actors in space continues to increase, there will undoubtedly be a corresponding increase in the laws applicable to those activities and actors. Despite the fact that international space law did not necessarily contemplate (and some argue is ill-equipped to address) the transformation of space activities that has taken place in recent years, States have largely shown little appetite for changing the existing international legal regime. Instead, States have addressed these evolutionary activities through the use of domestic law, as evidenced by a steady increase in the number of States that have implemented, and expanded upon, domestic legal regimes applicable to space and space-related activities. These States are not creating law for the sake of doing so; they are creating law due to a sense of legal obligation and/or a perception of need. The creation of law based upon legal obligation is tied to States' commitments under international space law, specifically the concepts of State responsibility and authorization and continuing supervision, as contained in Article VI of the OST. This sense of obligation ties in with the perception of practical need, which may largely be driven by emerging actors and technological developments that have changed fundamental calculations about the role of space in national policies.[19]

[19]For many States, the practical (from a purely domestic perspective) need to regulate space activities will coincide with the State's legal obligations under international space law, such that development of a domestic regime will serve both ends. However, for some States, international legal obligations may be the only driving factor, as those States may prefer to impose only the minimal regulatory regime necessary to meet the international obligations. Finally, for States that are not Party to the OST (and do not believe Article VI to be reflective of customary international law), it will only be the domestic perception of practical need that drives the development (or lack of development) of domestic law.

Space is risky and expensive, but the increasing access to space, the growing utility of space-based technologies, and the importance of space as an economic domain is making the risks more palatable, with more investors looking past the inherent risks to the potential rewards of being on the forefront of space development. These factors have contributed to an environment where more and more actors are accepting risk as part of entry into a potentially very profitable and dynamic domain. While many of the inherent risks are based on the physical properties of the domain, the novel nature of the technologies, and the financial commitment required, in some cases persistent legal risk underlies all others.

One facet of this legal risk arises when technological developments precede the law, resulting in potentially unregulated technologies that may force States to implement new laws or regulations, if for no other reason than to ensure the State is fulfilling its international obligations with respect to activities for which it may be held responsible.[20]International space law, particularly Article VI of the OST, gives States Parties an incentive, which is fairly unique in international law, to broadly regulate space activities – States can themselves be held responsible for such activities. Understanding the international legal framework is based on State responsibility, for both State and non-State space activities, is critical to understanding the heightened interest States may have in regulating such activities. Additionally, depending on how a State interprets its Article VI obligation to provide "authorization and continuing supervision," it may see itself as being legally required to provide some sort of regulatory framework for any space activity or technology that may conceivably fall within its jurisdiction.[21]

Consequently, those engaged in the development of innovative space technologies or activities should, especially in States with well-developed (but restrictive) domestic space law regimes, be cognizant of the fact their activities may trigger additional regulation, to include the associated risk that such regulations may effectively put an end to the activity or technology in question. In some legal systems, any activity or technology that is not explicitly authorized may be presumptively prohibited, and the legal requirements for regulatory approval may be high. Similarly, some States, when confronted with a new, unregulated technology, may regulate it in such as way that the utility of the technology is significantly decreased (whether intentionally or due to a lack of, or limited, understanding of the technology and/or its potential utility). As a result, the need to gain legal authorization for new activities or technologies may represent significant bar to

[20]In accordance with Article VI, the "appropriate State Party" is obligated to provide "authorization and continuing supervision" for the activities of non-governmental entities in space. Article VI also establishes that States Party "bear international responsibility for national activities in outer space ... whether such activities are carried on by governmental agencies or by non-governmental entities, and for assuring that national activities are carried out in conformity with" the OST.

[21]While in some instances legal arguments may be made regarding which State is the "appropriate State Party" to the OST to authorize and continually supervise a given activity, any State that has nationals engaged in space activities, or that has such activities taking place from or within its jurisdiction, certainly is likely to be deemed appropriate under Article VI.

innovation, especially if investors are wary of backing efforts which may never receive the legal authorization needed to make the effort viable or (perhaps equally problematic) if the efforts are authorized, but subject to a plethora of restrictions which significantly constrain the potential reach or market potential of the innovation.

While technological development may portend risks associated with the imposition of previously non-existent laws and regulations directed toward that technology, there is an inverse legal risk defined by the extent to which existing law itself constrains the development of new technologies. A most obvious (and likely intentional) example of this would be a case where a State implements preemptive rules in an attempt to steer, or outright prohibit, certain technological developments. In such an environment, there would be little incentive to incur the cost of working toward a technology, if the technology (or practical use of the technology) were limited or prohibited. In these cases, the risk of pursuing a prohibited technology is obvious. However, there may be other cases where existing constraints are perhaps not directly applicable to an emerging technology or space activity, leading to gray areas of law, where a nuanced understanding of the purpose of the law and regulatory/legislative history (along with an equally astute understanding of the novel technology and/or activity in question) may be the difference between establishing a legal basis for pushing the innovation forward and assuming it falls within the boundaries of existing constraints. Thus, an informed understanding of the law may effectively negate or mitigate risk to acceptable levels, allowing law to be used as an enabling force.

Law can provide certainty and that certainty may be enabling or constraining, depending on the nature and intent of the law. If the intent of the law is clearly to prohibit a particular technology or activity, those wishing to undertake activities or innovate in that space are likely facing an uphill battle, but a solid understanding of the basis of the prohibitory law and/or policy is certainly critical to any attempt to change the policy. The intent of the law, however, may not always be clear, especially when it comes to emerging technologies. While law is often driven by, and reflective of, policy, in cases of emerging technologies or innovative space activities, it may be the case that neither law nor policy actually contemplated the particular technology or activity. In such situations, understanding the history of the policy and/or the rationale for the law (or lack of law) becomes increasingly important, especially in those cases where neither a clear prohibition nor a clear authorization is evident. Law or policy which is not, on its face, intended to be a prohibition, should not be assumed to be one, especially in areas such as space activities and technology, where it is entirely possible the activity or technology at issue was simply never contemplated at the time the law or policy was put in place.

5.5 Conclusion

Like every domain in which human activities take place, space is subject to multilayered legal regimes, the complexity of which varies from State to State. The rules set by international space law are not, as is often the case, necessarily the same for all States, thus introducing differences in the rules (or at least some aspects of the rules) at a foundational level. The rules of the domain vary even more widely when domestic law and regulation is factored in, ultimately leading to vast disparities in the range of permitted and prohibited activities from State to State. Adding emerging technologies or innovative activities in the space domain into the equation only increases the complexity. In order to determine what a given domestic law or policy allows (particularly in the absence of a specific authorization or prohibition), it is necessary to have a sophisticated understanding of both the domestic law itself, as well as any international legal underpinnings of that law. An additional layer of complexity is added by the policy goals of the State in question, especially with respect the ways in which policy has influenced the development of domestic law, the extent to which that law is being used as a policy tool, and whether the particular law is aimed at satisfying domestic or international policy goals (or both). As emerging capabilities push the boundaries of what is technologically possible, and innovative activities challenge perceptions regarding traditional notions of the utility of the space domain, law becomes an enabling tool that will be most effectively wielded by those with a solid understanding of all aspects of the operative rules of the space domain.

References

Agreement Governing the Activities of States on the Moon and Other Celestial Bodies, 1363 UNTS 3, 18 December 1979 (entered into force 11 July 1984) [Moon Agreement].

Agreement on the Rescue of Astronauts, the Return of Astronauts and the Return of Objects Launched into Outer Space, 672 UNTS 119, 22 April 1968 (entered into force 3 December 1968) [Rescue and Return Agreement].

Convention on International Liability for Damage Caused by Space Objects, 961 UNTS 187, 29 March 1972 (entered into force 1 September 1972) [Liability Convention].

Convention on Registration of Objects Launched Into Outer Space, 1023 UNTS 15, 14 January 1975 (entered into force 15 September 1976) [Registration Convention].

Declaration of Legal Principles Governing the Activities of States in the Exploration and Uses of Outer Space, UNGA Res 1962 (XVIII), 13 Dec 1963.

Innovation, Science and Economic Development Canada. (2018). *Canadian table of frequency allocations* (2018th ed.*)*. https://www.ic.gc.ca/eic/site/smt-gst.nsf/vwapj/SMSE-07-18-CTFA-2018.pdf/file/SMSE-07-18-CTFA-2018.pdf

International Regulations in Force, 47 CFR §2.100.

International Telecommunication Union. (2020). *List of member states*. https://www.itu.int/online/mm/scripts/gensel8.

McDougal, S. (1963). The Emerging Customary Law of Space. *Faculty Scholarship Series, 2609*. http://digitalcommons.law.yale.edu/fss_papers/2609.

Question of the peaceful use of Outer Space, UNGA RES 1348 (XIII), 13 Dec 1958.

Treaty on Principles Governing the Activities of States in the Exploration and Use of Outer Space, including the Moon and Other Celestial Bodies, 610 UNTS 205, 27 January 1967 (entered into force 10 October 1967) [Outer Space Treaty].

UN COPOUS, Status of International Agreements Relating to Activities in Outer Space as at 1 January 2019, A/AC.105/C.2/2019/CRP.3 (1 April 2019).

UNOOSA. (2017, May). *International space law: United Nations instruments.* ST/SPACE/61/REV.2. https://www.unoosa.org/res/oosadoc/data/documents/2017/stspace/stspace61rev_2_0_html/V1605998-ENGLISH.pdf.

UNOOSA. (2020). *National space law collection.* https://www.unoosa.org/oosa/en/ourwork/spacelaw/nationalspacelaw/index.html

Chapter 6
To the Moon and Beyond: The Artemis Accords and the Evolution of Space Law

Melissa de Zwart

Abstract In 2020 NASA announced a bold plan to return humans to the Moon by 2024 and to use the Moon as a base for crewed missions to Mars. The vision and scope underpinning that plan represents a challenge not only to technological and scientific capability but also to existing understandings of international space law and invites a more detailed examination of key provisions of the UN Space Treaties. The Artemis project will rely significantly upon the involvement of commercial space operators and other states are invited through their respective space agencies to become part of the project. Involvement is however, contingent upon acceptance of and adherence to the 'Artemis Accords' reflecting 'a shared vision for principles, grounded in the Outer Space Treaty of 1967, to create a safe and transparent environment which facilitates exploration, science, and commercial activities for all of humanity to enjoy'. This chapter will consider the principles underpinning the Artemis Accords in the context of international space law and address the particular points of tension generated by their terms, notably the rules applicable to space resource utilisation and questions of territoriality, through the application of the contested principle of 'zones'. It concludes that merely tabling the Artemis Accords for consideration makes a valuable contribution to a much needed step forward in international space law.

6.1 Introduction

Space has become a renewed focus of both military and commercial operations in recent years. As the cost of launch and payload decreases, more nations recognise the opportunities presented by access to space, a domain that international law contemplated would only ever be the province of a very small number of space-faring Cold War superpowers and certainly only accessible by government

M. de Zwart (✉)
Adelaide Law School, The University of Adelaide, Adelaide, SA, Australia
e-mail: melissa.dezwart@adelaide.edu.au

M. de Zwart, S. Henderson (eds.), *Commercial and Military Uses of Outer Space*, Issues in Space, https://doi.org/10.1007/978-981-15-8924-9_6

operators. That context has drastically changed with the development of new fuels, materials, launch providers and purposes.

There are five major outer space treaties: primarily the *Treaty on Principles Governing the Activities of States in the Exploration and Use of Outer Space, including the Moon and Other Celestial Bodies* ('Outer Space Treaty'); as augmented by (in both date order and descending number of ratifications) the Rescue and Return Agreement (1968); the Liability Convention (1972); the Registration Convention (1975) and the Moon Agreement (1979) (collectively the 'UN Space Treaties'). The Outer Space Treaty, as at July 2020 has 110 states parties, with another 23 signatories in the process of completing ratification (COPUOS 2020). The Outer Space Treaty is therefore regarded as the overarching constitution of outer space, with several of its provisions also regarded as representing customary international law (He 1997, 53; Schmitt 2006, 99). However, in modern day space operations its provisions leave many gaps and grey areas which are not addressed in its provisions, nor in those of the other space treaties.[1]

Given the Cold War origins of the Outer Space Treaty, and the fact that only nation states were regarded as being capable of undertaking space activities, the reach and scope of the Outer Space Treaty provides little guidance with respect to modern day military, civilian and commercial uses of space. Its provisions were intended to maintain balance and transparency to ensure continued access to space by the Cold War superpowers, being the United States and the Soviet Union (de Zwart 2019, 339–343). Later treaties reflect an interest by non-spacefaring and developing countries in being able to access the growing benefits of access to the space environment. Considerable uncertainty now arises regarding the application of the existing international space law treaties to the modern uses of space.

Whilst there have been several attempts to fill these gaps with various 'soft law' instruments, including the UN sponsored program of Transparency and Confidence Building Measures, which issued its Report in 2013 (Report of the Group of Governmental Experts 2013); the European Code of Conduct for Activities in Outer Space, formally proposed in 2008; the Draft Treaty on the Prevention of the Placement of Weapons in Outer Space, the Threat or Use of Force Against Outer Space Objects, sponsored by Russia and China and tabled in the Committee on Disarmament in 2008 and 2014; and the inclusion of the Prevention of an Arms Race in Outer Space (PAROS) as a standing item on the agenda of the Conference on Disarmament since 1982, as well as a revised UNGA PAROS process in 2019 (Meyer 2016; Meyer 2018; Beard 2017) none of these processes has achieved consensus. It seems that the age of broad multilateral treaties on space law is long past.

This stalemate comes at a time when activity in space is increasing exponentially, and, particularly, when key activities in space are being undertaken by commercial providers. As will be discussed below, Article VI of the Outer Space Treaty

[1]See further, Trepczynski 'Rules of the Domain: The Impact of Law on the Evolution of Space Activities and Technologies' in this volume.

obliges States Parties to that Treaty to 'bear international responsibility for national activities in outer space... whether such activities are carried on by governmental agencies or by non-governmental entities'. This means that states are under increasing pressure to develop clear laws and guidelines to regulate New Space activities.

This chapter will consider the proposed NASA Artemis project to return to the Moon, and the rules being prescribed around participation in that project, in the context of the broader scope of the Space Treaties. It will address the key points of tension, being the extraction and use of resources, and the proposed use of safety zones around outer space activities. It concludes that whilst these articulated principles provide challenges to existing international space law, they provide an important starting point for a much needed clarification of the international law applicable to such activities at a time of vital need.

6.2 Return to the Moon

Recent years have seen a renewed interest in a return to the Moon. Both China and Israel dispatched lunar missions in 2019 (von der Dunk 2020), with the Chinese landing of a lunar rover in January and the Israeli mission crash landing and potentially distributing DNA samples and tardigardes on the Moon's surface in April (Oberhaus 2019). As with the Israeli mission, which would have been the first private moon landing, many of these achievements are reliant upon commercial space operators. Notable also is the successful crewed mission to the International Space Station undertaken by SpaceX launched on 30 May 2020, the first launch of American astronauts from US soil since the demise of the Space Shuttle programme in 2011. It is unsurprising therefore that NASA has announced that its return to the Moon will involve commercial and international partners.

Whilst these activities are conducted by private corporations, under the provisions of the Outer Space Treaty, the relevant state bears international responsibility for those activities, including authorisation and continuing supervision (Article VI) as well as international liability for damage caused by its space object (Article VII). The OST anticipates that states will ensure conformity with the provisions of the OST, most likely through the enactment of domestic space laws. However, these provisions leave many questions regarding how national space agencies, let alone private space corporations, may work together under the auspices of the relevant international law.

It should be noted that there is some precedent in this area. The most successful international space collaboration to date, the International Space Station (ISS), is operated pursuant to the International Space Station Intergovernmental Agreement (IGA) (Yakovenko 1999). That Agreement addresses many questions left unclear by international space law, such as ownership of intellectual property created within the ISS. Its provisions are supported by four Memoranda of Understanding between NASA, the European Space Agency (ESA), the Canadian Space Agency (CSA),

Roscosmos and the Japanese Space Agency (JAXA). In fact, it is anticipated that the IGA will be extended to apply to the Lunar Gateway project, being developed in conjunction with the Artemis project (Foust 2020). Hence there is hope that states, through their relevant agencies may be able to advance the resolution of key legal issues applicable to future uses of space through the conclusion of MOUs and agency to agency agreements.

The next section will address the principles underpinning the proposed Artemis Accords and place them in the context of the relevant international space law. It will assess their applicability and value with respect to the above conclusion that the time of multilateral treaties has finished and that the way forward to collaborative, rather than contested, uses of space will need to be found in multiparty agreements that also involve the private sector.

6.3 The Artemis Accords

NASA Administrator Jim Bridenstine released the Artemis Accords as a set of high-level principles on 15 May 2020, following reports in early May that the 'Trump administration is drafting a legal blueprint for mining on the moon under a new U.S.-sponsored international agreement called the Artemis Accords' (Roulette 2020). At that stage, key contentious issues had already been identified as the extraction and use of space resources and the development and enforcement of keep out zones on the Moon (Foust 2020).

As Borgen points out, the groundwork had been laid out in the Executive Order signed by President Trump on 6 April 2020 on 'Encouraging International Support for the Recovery and Use of Space Resources' (Borgen 2020). Critically, the Executive Order stated:

> Americans should have the right to engage in commercial exploration, recovery, and use of resources in outer space, consistent with applicable law. Outer space is a legally and physically unique domain of human activity, and the United States does not view it as a global commons. Accordingly, it shall be the policy of the United States to encourage international support for the public and private recovery and use of resources in outer space, consistent with applicable law. (Executive Order 2020)

That Executive Order affirmed the principles which had been set out in the US Commercial Space Launch Competitiveness Act 2015. Title IV of that Act, entitled 'Space Resource Exploration and Utilization' provided three key provisions: (1) The recognition of the rights of a US citizen 'engaged in commercial recovery of an asteroid resource or a space resource' to the resource so obtained; (2) A mandate for a US regime which required authorisation and supervision of such activities in conformity with US obligations under international law; and (3) A requirement for the President to promote and facilitate through appropriate Federal agencies, commercial space resources exploration and extraction and to promote the rights of US citizens to engage in such activities in accordance with international obligations (von der Dunk 2017).

The 2020 Executive Order however goes even further by explicitly repudiating the Moon Agreement, stating: 'US does not consider the Moon Agreement to be an effective or necessary instrument to guide nation states regarding the promotion of commercial participation in the long-term exploration, scientific discovery, and use of the Moon, Mars, or other celestial bodies.' Further, it explicitly concluded that the Moon Agreement does not reflect customary international law. The Moon Agreement stands in stark contrast to the Outer Space Treaty in terms of adoption. As at July 2020, it has only 18 ratifications and four signatories (COPUOS 2020).

The Executive Order reflected that: 'Successful long-term exploration and scientific discovery of the Moon, Mars, and other celestial bodies will require partnership with commercial entities to recover and use resources, including water and certain minerals, in outer space.' Importantly therefore (and acting within the ambit of Title IV) the Executive Order foreshadowed the need for some alternative road rules for the commercial use of space, setting the stage for the release of the Artemis Accords: 'With numerous countries and private sector players conducting missions and operations in cislunar space, it's critical to establish a common set of principles to govern the civil exploration and use of outer space.'

The Artemis Accords are articulated as a series of principles that must be agreed to by any state that wants to participate in the Artemis programme. In other words, other national space agencies that join NASA in the Artemis program will do so by executing:

> bilateral Artemis Accords agreements, which will describe a shared vision for principles, grounded in the Outer Space Treaty of 1967, to create a safe and transparent environment which facilitates exploration, science, and commercial activities for all of humanity to enjoy.

The approach of the Artemis Accords, using bilateral agreements to determine the rules of the joint venture, clearly build upon the success of the ISS IGA and it is likely that the first partners in the Artemis Accords venture will be the ISS partners, namely Canada, Europe, Japan and Russia. Other possible partners include Taiwan, Israel, India, South Korea, Australia and the United Arab Emirates (Whittington 2020).

The Artemis Accords consist of ten principles, although it is envisaged that these principles will be further developed in the terms of any pursuant bilateral agreement. Each of these principles will be addressed in turn below. Commentators (Foust 2020; Johnson 2020) have observed that the principles fall roughly into two extremes: existing norms that are broadly repetitive of the provisions of the Outer Space Treaty and those that extend well beyond its provisions and attempt to establish new norms of behaviour. This section will identify the categorisation of each principle in turn, but as has already been identified, the new norms address rights of extraction and use of space resources and the creation and observation of safety zones. Falling somewhere between these two ends of the spectrum are principles dealing with the creation of international standards to ensure interoperability of space systems, the protection of lunar heritage and agreement to adhere to the UN Space Debris Mitigation Guidelines.

6.3.1 Peaceful Purposes

> International cooperation on Artemis is intended not only to bolster space exploration but to enhance peaceful relationships between nations.
>
> Therefore, at the core of the Artemis Accords is the requirement that all activities will be conducted for peaceful purposes, per the tenets of the Outer Space Treaty.

The term 'peaceful purposes' appears four times in the Outer Space Treaty: twice in the preamble: 'Recognizing the common interest of all mankind in the progress of the exploration and use of outer space for peaceful purposes.' and 'Desiring to contribute to broad international cooperation in the scientific as well as the legal aspects of the exploration and use of outer space for peaceful purposes'; and twice in the body of Article IV, which reads, in part: 'The Moon and other celestial bodies shall be used by all States Parties to the Treaty exclusively for peaceful purposes. The establishment of military bases, installations and fortifications, the testing of any type of weapons and the conduct of military manoeuvres on celestial bodies shall be forbidden. The use of military personnel for scientific research or for any other peaceful purposes shall not be prohibited. The use of any equipment or facility necessary for peaceful exploration of the Moon and other celestial bodies shall also not be prohibited.'

Whilst the concept of 'peaceful purposes' is accepted as fundamental to international space law, its scope and meaning remain hotly contested. The term first appeared in the UN General Assembly (UNGA) Resolution 1148 of 1957[2]followed by UNGA Resolution 1348 (XIII) in 1958[3]which concerned 'Questions of the Peaceful Uses of Outer Space' and established the Committee on the Peaceful Uses of Outer Space (UNCOPUOS) as an ad hoc committee of the UNGA. UNCOPUOS was established as a permanent committee of the UNGA on 12 December 1959.[4]UNCOPUOS is mandated to deal only with the peaceful uses of outer space, with matters relating to military uses of outer space dealt with by the UN Conference on Disarmament. Therefore, despite having a long tradition at the heart of the negotiation and formulation of the Outer Space Treaty, its precise meaning is open to conjecture. It clearly is not intended to exclude all military uses of outer space, as militaries of various nations have been involved in space activities since the dawn of the space age. Further, there is extensive use of space by militaries for communication, surveillance and navigation, and the prohibitions on weaponization have not been interpreted as preventing the transit of weapons of mass destruction through space. State practice firmly supports a range of military uses of space as permitted within the meaning of peaceful purposes (de Zwart 2019, 344–345).

It should also be noted that the Moon Agreement (Article 3) also states that the Moon shall be used 'exclusively for peaceful purposes'. It however goes further by

[2]UNGA Resolution 1148 (XII) 14 November 1957.

[3]UNGA Resolution 1348 (XIII) 13 December 1958.

[4]UNGA Resolution 1472 (XIV) 12 December 1959.

prohibiting any 'threat or use of force or any other hostile act or threat of a hostile act on the Moon.' As with Article IV of the OST it affirms that 'the establishment of military bases, installations and fortifications, the testing of any type of weapon and the conduct of military manoeuvres on the Moon shall be forbidden. The use of military personnel for scientific research or for any others peaceful purposes shall not be prohibited.'

Thus there is little either contentious or concrete in this first principle. It merely endorses the long standing principle that space must be used for peaceful purposes, echoing the words of President Eisenhower to the UNGA in 1960 that it should 'press forward with a programme of international co-operation for constructive peaceful uses of outer space under the United Nations.' (Schrogl and Neumann 2009)

6.3.2 Transparency

> Transparency is a key principle for responsible civil space exploration and NASA has always taken care to publicly describe its policies and plans.
>
> Artemis Accords partner nations will be required to uphold this principle by publicly describing their own policies and plans in a transparent manner.

Clearly transparency is an essential element of multilateral cooperation. Issues of transparency and rights of inspection and verification were also key matters of contention during the early days of space exploration and the negotiation of the Outer Space Treaty. Russia, in particular, was keen to retain an obligation regarding verification of space objects, and the appearance of a commitment to such a principle appears in US President Eisenhower's 1960 speech to the UNGA. However, the practicalities of retaining a technological advantage through the maintenance of secrecy prevailed. Despite this complexity, principles related to information sharing and the desirability of transparency are recognised in several provisions of the OST, including Article IX (requests for consultation regarding activity or experiment potentially causing harmful interference), Article X (requests for an opportunity to observe the flight of objects launched by another state), Article XI (obligation to inform the UN, the public and international scientific community regarding the nature, conduct, locations and results of activities in outer space), and Article XII (requiring all stations, installation, equipment and space vehicles on the Moon and other celestial bodies to be open to representatives of other States Parties on a basis of reciprocity, subject to reasonable notice and appropriate consultations). However, despite these provisions, there is little evidence of states undertaking open information sharing, hence the need for Artemis partners to agree to such a practice, as it will be vital between project participants.

6.3.3 Interoperability

> Interoperability of systems is critical to ensure safe and robust space exploration.
> Therefore, the Artemis Accords call for partner nations to utilize open international standards, develop new standards when necessary, and strive to support interoperability to the greatest extent practical.

This principle is a novel development and one necessitated by commercial realities. The Artemis project contemplates that states will participate through their national space agencies, but that specific tasks and activities may be undertaken by commercial providers, such as the SpaceX mission to the ISS. The engagement of a broad range of commercial entities may require the use of Commercial Off the Shelf (COTS) technology to reduce the cost of manufacture, use and replacement. Hence, the requirement for standards development and interoperability is a new concept for space law and regulation, relevant to safety, efficiency and cost.

6.3.4 Emergency Assistance

> Providing emergency assistance to those in need is a cornerstone of any responsible civil space program.
> Therefore, the Artemis Accords reaffirm NASA's and partner nations' commitments to the Agreement on the Rescue of Astronauts, the Return of Astronauts and the Return of Objects Launched into Outer Space.
> Additionally, under the Accords, NASA and partner nations commit to taking all reasonable steps possible to render assistance to astronauts in distress.

As stated in the principle, astronauts have long held a special status in the Space Treaties. Article V OST provides that States Parties must regard astronauts as 'envoys of mankind' and 'render them all possible assistance in the event of accident, distress, or emergency landing'. Astronauts must be 'safely and promptly returned to the State of registry of their space vehicle.' Astronauts of one state must also render assistance to the astronauts of other states. These provisions are further enhanced and elaborated by the Rescue and Return Agreement.

A complicating factor is the scope of definition of who is an astronaut. The Rescue and Return Agreement uses the language of 'personnel of a spacecraft'. This also reflects the use of the terminology of personnel in Article VIII OST. Article 10 of the Moon Agreement attempts to reconcile both terms by declaring that any person on the Moon shall be both an astronaut within the meaning of Article V OST and personnel of a spacecraft under the Rescue and Return Agreement. More contentious will be the issue in practice as to whether all people working under the auspices of the Artemis Accords will be considered astronauts. Private space operators such as SpaceX, Blue Origin and Virgin Galactic currently all use the terminology of 'astronauts', however there is also reference to commercial crew and other roles. NASA uses terms to distinguish between crew, astronauts and space flight participants.[5] Thus while the principle of assistance is clear, the issue of who may be entitled to such assistance will require further development.

6.3.5 *Registration of Space Objects*

> Registration is at the very core of creating a safe and sustainable environment in space to conduct public and private activities. Without proper registration, coordination to avoid harmful interference cannot take place.
>
> The Artemis Accords reinforces the critical nature of registration and urges any partner which isn't already a member of the Registration Convention to join as soon as possible.

As is clearly articulated in this principle, registration is a fundamental element of the responsible use of the space environment and is embodied in Article VIII OST and the Registration Convention 1975.

6.3.6 *Release of Scientific Data*

> NASA has always been committed to the timely, full, and open sharing of scientific data.
>
> Artemis Accords partners will agree to follow NASA's example, releasing their scientific data publicly to ensure that the entire world can benefit from the Artemis journey of exploration and discovery.

Certainly Article XI OST obliges the sharing of information with the UN, the public and the scientific community. Again, whilst this imposes an obligation on the state party, it will remain to be seen how issues of commercial in confidence and confidential information will be handled under the Artemis project. The IGA does contain specific provisions dealing with management of information and this may provide a template for further development of this issue. However, there is clearly a need to clarify intellectual property and rights of commercialisation concepts in this context (de Zwart 2020).

6.3.7 *Protecting Heritage*

> Protecting historic sites and artifacts will be just as important in space as it is here on Earth.
>
> Therefore, under Artemis Accords agreements, NASA and partner nations will commit to the protection of sites and artifacts with historic value.

Clearly this principle is directed to the Apollo landing site, as well as the places of and artifacts left behind or created as a consequence of other crewed and robotic

[5] United States Chapter 509- Commercial Space Launch Activities Act, §50902.

lunar missions (NASA 2011). Whilst the US retains ownership of the space objects, the impact created on the Moon surface and resulting site is not owned by any State.

There has been a significant increase in interest in the heritage value of Moon sites in recent years. As Gorman (this volume) discusses, the moon landing site presents unique difficulty as a heritage site, given its location off-Earth (meaning it is not subject to the World Heritage Convention) and the fact that Article II of the OST itself prevents claims of sovereignty over the Moon by any means, rendering it impossible for any State to include the site on its own national register of protected sites. Other initiatives include projects such as For All Moonkind, an NGO proposing to develop a convention to protect human heritage in space (For All Moonkind n.d).

6.3.8 Space Resources

> The ability to extract and utilize resources on the Moon, Mars, and asteroids will be critical to support safe and sustainable space exploration and development.
>
> The Artemis Accords reinforce that space resource extraction and utilization can and will be conducted under the auspices of the Outer Space Treaty, with specific emphasis on Articles II, VI, and XI.

This element of the Accords clearly builds on the US Commercial Space Launch Competitiveness Act discussed above and is the key element of the Accords.

The Moon Agreement, which the US has now explicitly rejected, asserted in Article 11 that the 'Moon and its natural resources are the common heritage of mankind'. It obliged States Parties to that Agreement to establish an international regime to 'govern the exploitation of the natural resources of the Moon as such exploitation is about to become feasible'. That regime would encompass the rational management of such resources, as well as 'an equitable sharing by all States Parties in the benefits derived from those resources, whereby the interests and needs of the developing countries, as well as the efforts of those countries which have contributed either directly or indirectly to the exploration of the Moon, shall be given special consideration.' One such attempt to develop a set of such guiding principles is the Hague International Space Resources Governance Working Group, which has developed and adopted the Building Blocks for the Development of an International Framework on Space Resource Activities (Hague International Space Resources Governance Working Group 2019).

The Artemis Accords reflects an explicit move away from a broad international regime, with a focus on explicit development of commercial resource extraction, as flagged in the US Commercial Space Launch Competitiveness Act discussed above.

6.3.9 *Deconfliction of Activities*

Avoiding harmful interference is an important principle of the Outer Space Treaty, which is implemented by the Artemis Accords.

Specifically, via the Artemis Accords, NASA and partner nations will provide public information regarding the location and general nature of operations which will inform the scale and scope of 'Safety Zones'.

Notification and coordination between partner nations to respect such safety zones will prevent harmful interference, implementing Article IX of the Outer Space Treaty and reinforcing the principle of due regard.

The concept of safety or keep out zones have been proposed as a means of dealing with a more congested space environment for some time. The ISS is protected by a 200 km 'keep out' zone. Further, keep out zones have been suggested by space scholars as a means of protecting satellites in orbit, as well as in the context of resource extraction. Safety zones have thus been justified on the basis of space law concepts of harmful interference and due regard. Mining zones may create a significant hazard for personnel working on or under the Moon's surface and the risks and complexities created by lunar dust have been well documented (David 2019). The Hague Space Resources Working Group, for example, provides that the international framework 'should permit States and international organizations responsible for space resource activities to establish a safety zone, or other area-based safety measure, around an area identified for a space resource activity as necessary to assure safety and to avoid any harmful interference' with that activity (Hague Space Resources Working Group 2019). Any such zones would need to take account of principles of free access under international law and the need for notice and consultation.

Zones however also raise the possibility of undermining the principle of non-appropriation in Article II of the Outer Space Treaty which states that 'Outer space, including the Moon and other celestial bodies, is not subject to national appropriation by claim of sovereignty, by means of use of occupation, or by any other means.' Therefore, any practical implementation of this principle would need to be consistent with Article II.

Commentators have considered how, in this context, zones could be both established and exercised. Johnson for example, argues that the concept of keep out zones are justified under the terms of the OST as their practical implementation provides some real context for the concept of due regard under Article IX (Johnson 2020). Zones as articulated by the Artemis Accords could clearly and practically establish how various uses of the Moon could co-exist where there is adequate co-operation in the creation and recognition of various usage zones. Further questions arise as to whether zones must be limited to physical or safety matters or could be extended to economic or exclusive 'keep out' zones. Johnson notes that 'it would be contrary to the Outer Space Treaty to create blanket "keep out" zones. Rather, these Artemis safety zones will have to be articulated carefully to partners, and to the international community, in a way that stresses that Artemis activities are not asserting property rights over the boundaries of such zones. Instead, they should be explained as the physical manifestation of due regard amongst Artemis partners.' (Johnson 2020) Others, such as O'Brien (2020) have flagged the consequent issues of enforcement of such zones, asking if, for example, the new US Space Force will be deployed to protect economic zones? All of these details remain to be worked through and critically, will require not only the agreement of the parties, but the compliance of third parties who are not part of the Artemis Accords, in order to be effective.

6.3.10 Orbital Debris and Spacecraft Disposal

> Preserving a safe and sustainable environment in space is critical for both public and private activities.
>
> Therefore, under the Artemis Accords, NASA and partner nations will agree to act in a manner that is consistent with the principles reflected in the Space Debris Mitigation Guidelines of the United Nations Committee on the Peaceful Uses of Outer Space.
>
> Moreover, NASA and partner nations will agree to plan for the mitigation of orbital debris, including the safe, timely, and efficient passivation and disposal of spacecraft at the end of their missions.

These provisions seem to be consistent with the global recognition of the increasingly congested nature of space and the need to ensure that space debris needs to be managed and reduced to ensure that space remains accessible to all.

6.4 Evolution of Space Law?

Clearly, the intention behind the Executive Order and the Artemis Accords is to influence a shift in the development of international space law both via influencing, through rhetoric and activity, the interpretation of existing space law treaties and more controversially through the development of new customary international law through persistent and dispersed state practice, whilst at the same time purporting to remain within the overarching goals of the Outer Space Treaties. As Fidler observes: 'the United States is using existing governance regimes to advance space activities rather than pursue revision of the Outer Space Treaty or negotiate a new agreement.' (Fidler 2020) It is clear that whilst, as Fidler and others have noted, this reflects the Trump administration's rejection of multilateral treaties, it is fair to say that there has been very little appetite generally for another multilateral space treaty for many years.

However, it is also true that '[t]hrough the accords, the United States is using its influence as the predominant spacefaring nation to strengthen [its version of] international space law' (Fidler 2020). In fact, the United States administration is explicit on this point. As the principles are included in bilateral agreements, practice and adherence to their terms will influence and develop space law. There is also the issue that whilst parties to Artemis Accords might agree principles between themselves, for example to observe certain zones, there is no obligation on third parties to respect such agreement or zones (von der Dunk 2020). Of course, not all states undertaking missions to the Moon will do so in partnership with the US. Despite this, NASA's Mike Gold, Acting Associate Administrator for International and Interagency Relations, stated in his briefing to foreign press in May 2020: 'I hope that positive precedent that the principles of the Artemis Accords set will be adopted even by those who wouldn't necessarily be a member of an Artemis Accords agreement.' (Gold 2020)

Roscosmos has already articulated its position that it is open to discussions regarding participation by Russia in the Artemis program and consequent agreement to the Artemis Accords. This is at odds with the initial reaction of the Head of Roscosmos, Dmitry Rogozin, who likened the proposed Artemis Accords as akin to the US invasion of Iraq, in a series of Tweets (Sheldon 2020). However, subsequent comments by both Russian and US space agencies have indicated a willingness to work together, particularly with respect to the Lunar Gateway project.

It is clear that the release of the Artemis Accords is a much needed step forward in the formulation of principles that will apply to an increased human engagement with the exploration and use of space. Whether one agrees or disagrees with the individual principles, noting that several of them are more akin to motherhood statements than radical calls to action, they have at the very least reignited a practical engagement with the question of what is permitted in space under international law and has placed the status and application of the Moon Agreement under the spotlight. As Schingler notes: the Accords 'raise a rich set of policy questions as we begin to take the law into new levels of resolution. This bold pursuit of uncharted territories is to be applauded, and yet, there is also the risk of diverging from 53 years of international law.' (Schingler 2020)

Johnson argues that, rather than being a land grab by the US, the Artemis Accords reflect a desire to make the Moon a co-operative zone, that the Accords seek to establish new norms that prioritise co-operation and relationship building, through principles such as interoperability, registration, transparency and data sharing. He argues that the operation of Article IV means that 'Peaceful activity on the Moon is not solely for the purposes of a conflict-free lunar landscape, but it is tied to relationship maintenance and building between nations on Earth. Concerns that the US Space Force will be operating on the Moon, are, in my view, unfounded, because of this strong, non-militarization clause, a pillar of space law.' (Johnson 2020).

6.5 Conclusions

Foundations have been laid for the development of new rules for the space domain as we move from the age of scientific discovery to the phase of commercial exploitation (Stimers 2020). As noted above however, a key sticking point at the moment in international space law is the Moon Agreement. In 2019, the Australian

Government, through the Australian Space Agency, entered into a partnership with NASA, allocating $150M for Australian businesses and researchers to join the NASA Moon to Mars initiative. However, Australia is one of the few States Parties to the Moon Agreement, and as has already been observed 'Australia would be obliged to withdraw from the Moon Treaty (sic) if it accepts an offer to join the Accords' (Whittington 2020).

Therefore, as the Artemis project moves forward, states may find themselves under increased pressure from commercial space operators who want to play a part in the new missions to the Moon and beyond. This will necessarily involve a reassessment by states regarding how new rules can be put in place to facilitate the involvement of commercial operators, as well as the undertaking of new activities, such as resource extraction and utilisation. This may also involve novel concepts such as zones.

It is clear that if missions are to push beyond the Moon to Mars and further, a transparent regime surrounding resource extraction will need to be developed and agreed upon. That agreement requires at minimum agreement from all of the space faring nations and, if the Moon Agreement model prevails, other states as well. The Artemis Accords may not end up being the final word on what commercial missions to the Moon will look like, but they reflect a reasoned and sensible attempt to move the dialogue on these issues along. It has been too long since a multilateral treaty could address all of the complexities of space missions and we now need to use this new platform and others like it to meaningfully and thoughtfully move forward into space.

References

Agreement among the Government of Canada, Governments of Member States of the European Space Agency, the Government of Japan, the Government of the Russian Federation, and the Government of the United States of America concerning Cooperation on the Civil International Space Station, 29 January 1998 (entered into force 28 March 2001) (IGA)

Agreement among the Government of the United States of America, Governments of Member States of the European Space Agency, the Government of Japan, and the Government of Canada on Cooperation in the Detailed Design, Development, Operation, and Utilization of the permanently Manned Civil Space Station, 29 September 1988) (entered into force 30 January 1992)

Agreement Governing the Activities of States on the Moon and Other Celestial Bodies, opened for signature 18 December 1979, 1363 UNTS 3 (entered into force 11 July 1984) ('Moon Agreement')

Agreement on the Rescue of Astronauts, the Return of Astronauts and the Return of Objects Launched into Outer Space, opened for signature 22 April 1968, 672 UNTS 119 (entered into force 3 December 1968) ('Rescue and Return Agreement')

Beard, J. (2017). Soft Law's Failure on the Horizon: The International Code of Conduct for Outer Space Activities. *University of Pennsylvania Journal of International Law, 38*(2), 335.

Borgen, C. (2020, May 8). 'The Artemis Accords: One Small Step for Space Law?'. *Opinio Juris (Online)*. http://opiniojuris.org/2020/05/08/the-artemis-accords-one-small-step-for-space-law/

Committee on the Peaceful Uses of Outer Space (COPUOS): Legal Subcommittee. (2020, January 1). Status of International Agreements relating to activities in outer space. https://www.unoosa.org/documents/pdf/spacelaw/treatystatus/TreatiesStatus-2020E.pdf

Convention on International Liability for Damage Caused by Space Objects, opened for signature 29 March 1972, 961 UNTS 187 (entered into force 1 September 1972) ('Liability Convention')

Convention on Registration of Objects Launched into Outer Space, opened for signature 14 January 1975, 1023 UNTS 15 (entered into force 15 September 1976) ('Registration Convention')

David, L. (2019 October 21). 'Moon Dust Could Be a Problem for Future Lunar Explorers'. Space.com *(online)*. https://www.space.com/moon-dust-problem-lunar-exploration.html

de Zwart, M. (2019). Outer Space. In W. H. Boothby (Ed.), *New technologies and the law in war and Peace* (p. 337). Cambridge: Cambridge University Press.

de Zwart, M. (2020). Intellectual property, innovation and new space technology. In G. Austin, A. Christie, A. Kenyon, & M. Richardson (Eds.), *Across intellectual property: Essays in honour of Sam Ricketson, Cambridge Intellectual Property and Information Law* (pp. 144–157). Cambridge: Cambridge University Press. https://doi.org/10.1017/9781108750066.014.

Fidler, D. P. (2020, June 2). The Artemis Accords and the next generation of outer space governance. *Council on Foreign Relations*. https://www.cfr.org/blog/artemis-accords-and-next-generation-outer-space-governance

For All Moonkind. https://www.forallmoonkind.org/

Foust, J. (2020, July 13). What's in name when it comes to an "accord"?. *The Space Review* (online). https://www.thespacereview.com/article/3987/1

Gold. (2020 May 15). Artemis Accords enabling international partnerships for lunar exploration. *US Department of State Foreign Press Centers Briefing*. https://www.state.gov/briefings-foreign-press-centers/artemis-accords-enabling-international-partnerships-for-lunar-exploration/

Hague International Space Resources Governance Working Group. (2019, November). *Building Blocks for the Development of an International Framework on Space Resource Activities.*

He, Q. (1997). The Outer Space Treaty in Perspective. 40 *Proceedings on the Law of Outer Space,* 51, 53.

Johnson, C. (2020). SpaceWatchGL feature: The space law context of the Artemis Accords: Part II. *SpaceWacth Global*. https://spacewatch.global/2020/05/spacewatchgl-feature-the-space-law-context-of-the-artemis-accords-part-2/

Meyer, P. (2016). Dark forces awaken: The prospects for comparative space security. *Nonproliferation Review, 23*(3–4), 495.

Meyer, P. (2018, December 11). Washington Starts a Space Spat at the United Nations. *Bulletin of the Atomic Scientists.*

National Astronautics and Space Administration. (2011, July 20). NASA's recommendations to space-faring entities: How to protect and preserve the historic and scientific value of U.S. Government Lunar Artifacts.

Oberhaus, D. (2019, May 8). *A Crashed Israeli Lunar Lander Spilled Tardigrades on the Moon.* https://www.wired.com/story/a-crashed-israeli-lunar-lander-spilled-tardigrades-on-the-moon/.

O'Brien, D. (2020, 29 June). The Artemis Accords: repeating the mistakes of the Age of Exploration. *The Space Review (online)*. https://www.thespacereview.com/article/3975/1

Roulette, J. (2020, May 6). Trump administration drafting 'Artemis Accords' pact for moon mining. *Science News(online)*. https://www.reuters.com/article/us-space-exploration-moon-mining-exclusi/exclusive-trump-administration-drafting-artemis-accords-pact-for-moon-mining-sources-idUSKBN22H2SB.

Schingler, J.K. (2020 June 8). Imagining safety zones: Implications and open questions. *The Space Review (online)*. https://www.thespacereview.com/article/3962/1.

Schmitt, M. (2006). International law and military operations in space. In A. von Bogdandy & R. Wolfrum (Eds.), *Max Planck yearbook of United Nations law* (Vol. 10, p. 99).

Schrogl, K. U., & Neumann, J. (2009). [Outer Space Treaty] Article IV. In S. Hobe, B. Schmidt-Tedd, & K.-U. Schrogl (Eds.), *Cologne commentary on space law volume 1: Outer space treaty.* Carl Heymanns: Cologne.

Sheldon, J. B. (2020) Lunapolitics: Russia's Roscosmos Signals Openness to US-proposed Artemis Accords. *Spacewatch Global*. https://spacewatch.global/2020/05/lunapolitics-russias-roscosmos-signals-openness-to-us-proposed-artemis-accords/.

Stimers, P. (2020, April 20). The President's space resources executive order: A step in the right direction. *The Space Review*. https://www.thespacereview.com/article/3925/1

Treaty on Principles Governing the Activities of States in the Exploration and Use of Outer Space, including the Moon and Other Celestial Bodies. (1967, January 27). 18 U.S.T. 2410, 610 U.N.T.S. 205 (entered into force 10 October 1967) ('Outer Space Treaty').

von der Dunk, F. (2017). Asteroid mining: International and national legal aspects. *Michigan State International Law Review, 26*(1), 83.

von der Dunk, F. (2020, June 2). The Artemis Accords and the law: Is the Moon "back in business"? *University of Auckland Blog*. https://www.thebigq.org/2020/06/02/the-artemis-accords-and-the-law-is-the-moon-back-in-business/

Whittington, M. (2020, April 26). How the United States plans to make space exploration pay. *The Hill*.

Yakovenko, A. (1999). The intergovernmental agreement on the International Space Station. *Space Policy, 15*(2), 79–86. ISSN 0265-9646. https://doi.org/10.1016/S0265-9646(99)00011-9.

Chapter 7
The Case for Space Arms Control

Laura Grego

Abstract A dangerous space arms race dynamic has been unfolding for decades but has begun accelerating in recent years. The ability to hold satellites at risk is becoming more sophisticated and is proliferating, both on the ground and in orbit. Absent correction, the space strategic environment will become more threatening and unpredictable and the space physical environment for all users may also degrade because of fast-moving space debris. States will spend more resources preparing for conflict and crises. Space arms control is urgently needed and can be achieved, given political will. It could help to provide a future in space that is more secure and more sustainable and can be used to help create a more safe and secure future on earth.

7.1 Introduction

On March 27, 2019, India launched a missile defense interceptor into space, intentionally obliterating one of its low-earth orbiting satellites with the force of impact (Ministry of External Affairs 2019). Prime Minister Modi declared that "Mission Shakti" had put India in an elite group of space powers, alluding to the fact that India indeed was only the fourth country to destroy its own satellite, after Russia, the United States, and China (PTI 2019). While the satellite had been destroyed at a low enough altitude to ensure that atmospheric drag could clear most of the detritus out of orbit within days or weeks, some debris was thrown up into much higher altitudes. That debris expected to stay in orbit for months or years, where it could collide with other satellites (Henry 2019).

Mission Shakti was a manifestation, visible to those with eyes on the sky, of a dangerous space arms race dynamic that has been unfolding rather slowly for decades but which has been recently picking up speed. The ability to hold satellites at risk is becoming more sophisticated and is proliferating, both on the ground

L. Grego (✉)
Union of Concerned Scientists, Cambridge, MA, USA
e-mail: lgrego@ucsusa.org

M. de Zwart, S. Henderson (eds.), *Commercial and Military Uses of Outer Space*, Issues in Space, https://doi.org/10.1007/978-981-15-8924-9_7

and in orbit (Secure World Foundation 2020; Harrison et al. 2020). In June and November of 2019, Russia conducted tests of a system reported to be its own ground based anti-satellite weapon (though the tests did not include actually destroying a satellite) and the US decried maneuvers by Russian satellites around sensitive US satellites (Secure World Foundation 2020; Hennigan 2020). France, having also been frustrated by a loitering Russian satellite around its own national security satellite, unveiled a new military space strategy and announced the right to an "active defense" of its satellites (Ministere des Armees 2020). The US' fiscal year 2020 defense plan included programs to build space-based missile defense interceptors and directed energy weapons, though they were eventually stripped from the budget by Congress. Upon the December 20, 2019 presidential signing of that defense budget, the United States established the Space Force as a separate military branch, giving responsibility for preparing for the US to establish "superiority" in space and to perform offensive and defensive space "control." (NDAA 2019)

This is unfolding against a backdrop with little diplomatic energy spent to constrain this dynamic. In fact, a few days after India's test, a two-year United Nations process to examine practical measures to prevent an arms race in outer space ended with failure to reach consensus on recommendations. Few alternative processes are available; the United Nations Conference on Disarmament remained in the deadlock it has been in for decades (Ministere des Armees 2020).

Absent correction, the space strategic environment will become more threatening and unpredictable. And the space physical environment for all users may also degrade because of fast-moving debris from anti-satellite tests, even if done with nominal care to produce the smallest amount of debris possible. States will spend more resources preparing for conflict and crises. And since commercial and other civil endeavors operate in the same physical space as national security space activities, they will also have to plan for a more unpredictable and dangerous environment. And this uncertainty will not be constrained to space. Space activities risk sparking or exacerbating terrestrial crises in ways that may be difficult to predict or manage.

What can be done about this? Is arms control for space necessary? Is it achievable? This chapter aims to provide evidence that thoughtful space arms control can help to provide a future in space that is more secure and more sustainable and can be used to help create a more safe and secure future on earth.

7.2 Is Arms Control Necessary? What Problem Is It Trying to Solve?

To understand whether arms control might be useful, one has to be clear about its goals. While the foundational theoretical work[1] by Schelling and Halperin (1961)

[1]The study and practice of arms control has decades of substantial literature.

defines arms control as "all the forms of military cooperation between potential enemies in the interest of reducing the likelihood of war, its scope and violence if it occurs, and the political and economic costs of being prepared for it", practitioners have in the decades since identified numerous specific goals that arms control might serve, including avoiding arms racing, curtailing incentives to strike first in an incipient crisis, and reducing the paths to inadvertent escalation of a crisis because of misperception (Schelling and Halperin 1961; Brooks 2020). The events of the last year illuminate three dangerous trends for which arms control might provide guardrails, including: an increasingly threatening space environment that exacerbates the risks of starting or escalating a crisis; an arms race dynamic in which space actors arm themselves with offensive weapons in space and aimed at space systems and defensive weapons to counter them; and the contamination of the space environment by debris from weapons tests and use. This section looks more closely at how these goals play out in the space environment before turning to what arms control for space might look like and how to get started.

7.2.1 *Reduce the Risk of Conflict and Provide Stability*

A number of characteristics of space operations make it ripe for providing pathways to unplanned or mistaken crisis.

Satellites are inherently fragile and difficult to protect, at least individually. Launch is expensive, costing thousands of dollars per kilogram, so satellites don't generally carry protective armor. Satellites are visible from the ground and move in repeated, predictable orbits. While satellites in geosynchronous orbits are provided some protection from earth-based threats by virtue of their distance (they orbit at approximately 36,000 kilometers altitude), satellites in the heavily-used low-earth orbits are relatively local, orbiting not more than 2000 kilometers from the earth's surface.

Satellites can be held at risk using technologies with a wide range of attributes, from temporary and reversible cyber attacks on command and control systems to permanently destructive attacks such as India's missile-borne interceptor; from unsophisticated jamming of communications channels to technically complex weapons that can stealthily approach a target satellite; from targeted attacks using high powered microwaves that stop a satellite's electronics from working but otherwise leave it intact to indiscriminate attacks such as a nuclear detonation in space that can disable dozens of satellites at once.

Satellites' vulnerability to interference is an acute concern because of their importance to states with technologically sophisticated national security infrastructure that depends on space. For a small but increasing number of sophisticated space actors, space services provide the capability to deploy military personnel globally and apply force more efficiently and effectively. Increasingly, these states view space as a sphere to be dominated to defend these advantages.

The perceived vulnerability of these valuable assets can lead to an incentive to strike first in an incipient crisis. If hostilities look probable, a state with substantial dependence on space may have an incentive to take military action first, before critical satellites it depends on are lost to their adversary's first strike. This dynamic is likely to be particularly acute if that adversary has demonstrated or fields significant anti-satellite capability. Even if both (or all) sides in a crisis don't plan to and prefer not to engage in armed conflict, this perceived vulnerability may pressure them to approach it closely anyway.

The nature of space operations may also create pathways to mistaken crisis escalation. Because of the remoteness of satellites, it can be difficult to get timely information about what is happening to a satellite and about the space environment around it, the "situational awareness" is imperfect and subject to delays. And so it may be difficult to understand whether a given satellite failure is due to natural causes, inadvertent interference, or intentional attack. In an increasingly threatening environment, where many actors field anti-satellite weapons, even during times of peace, operators may assume malicious intent. More to the point, if the costs of reacting too slowly in a crisis are perceived to be high, an incentive can be created to choose the worst-case interpretation of events, even if the information is incomplete or inconclusive.

Adding to the problem is that norms of behavior have not been well-established and so activities are ripe for misinterpretation. Because there is no history of a major conflict that includes space, and little high-level strategic discussion about space among the powers who are potential adversaries, states may lack a shared understanding of what kinds of actions the others consider out of bounds. What one state thinks is an appropriate reaction to a situation may seem out of proportion to another. For example, if an unfriendly state moves its satellite close to a potential adversary's sensitive satellite without permission or notification, that potential adversary may declare it has a right to anticipatory self-defense and target that maneuvering satellite.

7.2.2 Avoid an Arms Race

While space has for decades been essentially adjacent to conflict, playing an important, but supporting role, it has been moving to the center. Satellites have been providing services needed for increasingly sophisticated and global deployment of military force. Satellites provide services and data that states find critical for national security missions. These include intelligence, surveillance and reconnaissance; precision navigation and timing (PNT) signals; weather monitoring and prediction; global high-throughput and secure communications; monitoring of compliance with arms control agreements and early warning of threats such as ballistic missile launch.

States are increasing their investments in these capabilities. Ten years ago, 228 satellites were primarily military in nature, today more than twice as many (477) are in orbit (UCS Satellite Database 2010, 2020).[2] While the United States is still far and away the heaviest user of space for military purposes, other countries are doing more. For example, in 2009, the United States held about 70% of military space capability, Russia 12%, and China 5%. In 2019, the US share was 55%, Russia 14%, and China 19% (Munich Security Report 2020).

Increasingly, space users indicate they expect that those satellites will be threatened and are taking steps to protect those investments as well as to hold potential adversary satellites at risk.

In 2020, the United States, the heaviest user of space, released its new Defense Space Strategy, motivated by perceived challenges from China and Russia:

> ...China and Russia present the greatest strategic threat due to their development, testing, and deployment of counterspace capabilities and their associated military doctrine for employment in conflict extending to space. China and Russia each have weaponized space as a means to reduce U.S. and allied military effectiveness and challenge our freedom of operation in space (United States Department of Defense 2020).

The strategy's plans to "ensure space superiority" included establishing a new branch of the armed forces, the Space Force, and the Space Command as a unified combatant command.

As described in the introduction to this chapter, and detailed in nongovernmental and governmental assessments, several states have begun to develop and field anti-satellite weapons in increasing numbers and with increasing sophistication. The United States, Russia, China and India have destroyed satellites in weapons tests, and retain such capabilities. The United States fields dozens of such ground-based missiles that can target low-earth orbiting satellites with a homing vehicle in the guise of its ballistic missile defense system (Grego 2011). China has reportedly begun operationalizing a similar system and has tested a missile that could take homing vehicles to the vicinity of higher-altitude satellites (Secure World Foundation 2020). Because intercontinental-range missiles and satellites traverse the same regions of space with similar speeds, missile defenses designed to target intercontinental ballistic missiles are generally also capable anti-satellite weapons.

Satellites that can closely approach a potential adversary's satellite without its cooperation—rendezvous and proximity operations satellites– can be used for beneficial purposes, such as refueling or repair but also for less benign missions such as eavesdropping, spying on, or damaging the target satellite. This technology is fielded by the United States, Russia, and China and others. Ground-based systems that can target command and control systems using cyber attacks or jamming of the uplinked signals are likely widely held, as is the ability to interfere with earth observing satellite sensors using ground-based lasers.

At least one country, the United States, is proposing to put weapons in space to counter ballistic missiles, though no actual programs to develop the systems have been initiated. These plans have included both directed energy systems—lasers or neutral particle beams—or interceptor missiles. The speeds needed for

[2]Data includes all satellites fielded by militaries or with militaries as a prime partner.

such interceptors to reach launching ballistic missiles and disable them would also provide enough speed to bring those interceptors to geosynchronous orbits in a matter of hours, and so could be used as ASAT weapons against those more remote and protected satellites.

These states are establishing bureaucracies to develop anti-satellite weapons technologies to pursue advantages or to avoid a decisive disadvantage, to cut their potential adversaries' advantages derived from satellites, and counter-ASAT weapons to defend these satellites. States are creating strategies and doctrines to use these weapons to achieve their security goals and establish superiority over the space environment. While observers may disagree at how far down the path we have proceeded, the ingredients for an arms race are in place.

7.2.3 Reduce Harm to the Space Environment

Space is not solely or even primarily a military or national security environment. More than 2600 satellites operated by more than 75 countries orbit the earth, military satellites currently comprise around 15% of the use of space, and a flourishing commercial and civil space sector provide science and commerce and development benefits (UCS Satellite Database 2020; Munich Security Report 2020). The nonmilitary sector of space will come to dominate even more in the coming years, as planned large commercial constellations will come online, adding potentially tens of thousands of satellites.

The space sector currently is estimated to generate $350 billion a year and predicted to be more than $1 trillion by 2040 (Morgan Stanley 2019). Space technologies are critical to progress on the United Nations Sustainable Development Goals 2030, for example, by providing communication to remote regions to undergird education and telemedicine, by using satellite-derived weather and temperature data to support crop management and avert environmental degradation, and space-based monitoring to hold governments and corporations accountable and to support conflict resolution (UNOOSA 2020). And of course, some kinds of priceless scientific data can only be obtained by satellites orbiting above the earth's atmosphere, and future space exploration will necessarily have to traverse earth orbit on its way.

Satellites share the same physical space; low-earth orbits that are so useful for observing earth and geostationary orbits that permit global broadcast and communications are useful for militaries for the same reason they are useful to the civil sector. Trends that increase risks and hazards will affect all users of space.

A particular challenge for satellites is the increasing amount of space debris. Because of the high speed characteristic of objects in orbit, a collision with a 5 cm object imparts as much energy as being hit by a bus (Thompson 2015). Collisions with even a small piece of debris, the size of a marble, can be catastrophic for a satellite (and potentially create yet more debris to create additional hazards, in a cascade effect), but debris this small currently cannot be tracked while larger objects such as a defunct satellite can be tracked and potentially avoided.

Space debris has historically come from the detritus of launch and of operation in space, including nonfunctioning satellites, lost equipment, and spent missile stages, intact or exploded. The international community of space users has spent considerable effort in developing best practices guidelines to mitigate the production of debris and to slow the addition of debris from routine space operations (UNOOSA 2010). This has met with moderate success (ESA 2019).

The single event that generated the largest amount of persistent space debris was China's intentional destruction of its Fengyun-1C satellite in 2007. That ASAT demonstration created an estimated 300,000 pieces of debris larger than 1 cm, large enough to cause damage in a collision. However, only around 3300 of these pieces were large enough to be tracked (Hansen et al. 2015). Fengyun-1C was a moderately-sized satellite; the destruction of single large 10-ton satellite in low earth orbit could double or triple the amount of persistent and dangerous space debris in that environment (Wright 2007). If the target were an adversary's satellite, it would likely be functioning and in orbits near other satellites. A conflict that led to multiple large satellites being destroyed could have catastrophic effects on the ability to use space safely. While states with investments and aspirations in space will tend to be self-deterred from using these kinds of weapons, the accumulation and proliferation of this capability leaves open the possibility of their use. Especially since these weapons are predictably effective and the success of their use is easy to confirm compared to other types of ASAT weapons.

Having identified three key risks that space activities and conflict might generate, the next section will look at space arms control frameworks broadly and how they might serve to mitigate these risks.

7.3 Is Space Arms Control Valuable? Is it Achievable?

Good arms control provides benefits to its adherents, limits advantages accrued to cheaters, and is verifiable. Sensible constraints on behavior and limits on dangerous space weapons technologies could help to augment stability and curb the pathways to conflict, avoid an arms race, and help preserve the space environment.

Under a regime which limited the types of weapons that could be tested or deployed, the incipient arms race dynamic could be curtailed, and ideally more cooperative models for security could be implemented. Space actors would have less incentive to try to counter such anti-satellite and space-based weapons with their own, and would have less incentive to "use or lose" capabilities underpinned by vulnerable space assets.

In the absence of limits on anti-satellite and space weapons, space actors are increasingly seeking other ways to improve their security. This includes moving to a "resilient" posture[3] to prepare for conflict by increasing the robustness of individual satellites to interference, and building on-orbit and ground-based back up systems for critical capabilities provided by satellites. For example, states may field ground-based communication lines or air-based navigation and timing systems to make up for a loss of communications or PNT satellites. Satellite operators may need to plan for rapid reconstitution of their constellations by having on-orbit backups or an ability to launch replacements quickly and to add shielding from debris or electromagnetic interference. Arms control can help reduce the threat environment and reduce the cost and complexity of these preparations.

7.3.1 What Might Space Arms Control Look Like?

Arms control could take a number of forms, including legally- or politically-binding multilateral agreements or could start with unilateral declarations. It could include limits on hardware or limits on behavior. But to be most effective, it is likely to include limits both on the most dangerous technologies and limits on behaviors, such as the way that dual-use technologies are used.

An example of a limit on a particularly dangerous technology is constraint on direct ascent interceptors of the type that the US, China, and India have tested. This could include limits on the test programs, which might create some limits to their operational readiness; limits on geographical location, to limit the satellites that could be targeted synchronously; limits on the range of the interceptor's launching missile, to limit the kinds of satellites that could be targeted; or limits on the number of interceptors, including banning them entirely. It is possible to carve out exceptions that would allow states to retain some missile defense capability, while limiting the ASAT capability of the systems. Because these systems use missile launches, which are readily detectable by space-based infrared sensors, the testing of these systems is visible to observers and any subsequent debris (or loss of a satellite) is also visible using ground-based sensors. These mechanisms provide information about the existence and capability of these systems, and could comprise a central part of a verification regime.

If more confidence in compliance is required than can be provided by these passive observations, then more rigorous and intrusive instruments can be designed. Limits on the testing and use of these types of weapons provide a number of benefits, including limiting debris production from tests and possible use, and can curb the destabilizing influence of these powerful and destructive systems.

An example of behavioral limits that could contribute to confidence-building and address a potential source of misunderstanding and miscalculation are those gov-

---

[3]The 2011 US National Security Space Strategy identified resilience as a way to continue to function through an attack as well as a deterrence strategy, by denying an attacker the benefits of the attack. See (US Department of Defense and Office of Director of National Intelligence 2011).

erning the operation of satellites performing rendezvous and proximity operations. Carefully negotiated norms that establish how close and under what conditions a satellite may approach another one can help establish confidence about a potential adversary's intentions and provide a way to identify actions that are out of bounds and require more attention. Space actors could reserve their energy for solving real problems rather than seeing them where they don't exist or misinterpreting actions as a deliberate signal or a precursor to an attack.

Success in adopting discrete and incremental steps like the above can build confidence that the international community of space users can identify common concerns and find agreed solutions. A piecemeal approach certainly may not be enough and a comprehensive approach to arms control might be the most durable way to address these issues.

At present, conditions do not support such an approach. The major space powers not only do not agree on the solution, they do not clearly share a sense of the problem. For example, China and Russia's strong preference is for a treaty-based approach negotiated under United Nations auspices with the framework of preventing an arms race in space. They have proposed the "Treaty for the Prevention of the Placement of Weapons in Outer Space," a comprehensive ban on the deployment of space-based weapons and on threats of any kind against satellites.[4] The United States has stated that it does not believe there is an arms race in space and sees little value in this treaty, but has not proposed changes to the text nor suggested its own preferred legally-binding elements. Efforts to negotiate a comprehensive treaty are also confounded by the fact that the United Nations Conference on Disarmament, the venue designated for arms control negotiations, has been deadlocked for decades.

Under the Obama administration, the US' preferred approach was to start with a politically-binding (rather than legally-binding) set of agreed norms. The US supported the European Union's proposal for an International Code of Conduct for Outer Space Activities, which would establish "rules of the road" and general principles of responsible behavior in space (EEAS 2015). Efforts to advance the Code ended in 2015 without success. And no subsequent process to develop a comprehensive set of norms has met with success.

Another approach is to create consensus around a limited set of objectives and to build outward from there. One key norm in the Code was that signatories would refrain from intentional destruction of or damage to satellites. This simple form of arms control has found support outside of the Code, as other states have advanced ideas such as a moratorium on testing anti-satellite weapons and bans on debris-creating weapons.[5]

---

[4]Foundational documents and drafts of the draft treaty texts and U.S. responses are available at the United Nations Conference on Disarmament archive of documents. See https://www.unog.ch/80256EE600585943/(httpPages)/D4C4FE00A7302FB2C12575E4002DED85?OpenDocument

[5]For example, Canada proposed to the UN Conference on Disarmament several measures, including a moratorium on antisatellite tests, (see:https://undocs.org/CD/1815) and a pledge not to use weapons to damage or destroy satellites (see: https://undocs.org/CD/1865). Germany has

Whether the most promising path starts with negotiation of legally binding agreements or the establishment of norms, whether they focus on limiting technology or limiting behavior, a few key elements are likely to be necessary. These are discussed in the next section.

7.3.2 *What Elements Are Needed for Arms Control?*

7.3.2.1 Definitions

A common criticism about space arms control, such as the Russian and Chinese proposed treaty, is that it is difficult to define a "space weapon" (Beard 2017). This does not have to be a showstopper; rather than a definition of a few sentences, arms control treaties regularly include lengthy, detailed technical sections that serve to define the systems they are concerned with and a consultative process for resolving disputes.[6]

Advancing a specific definition for space weapons is not the scope of this chapter, but it is useful to understand broadly that space technologies are not so different as to be undefinable. Some space systems are unambiguously weapons, most satellites are unambiguously not usable as weapons, and the remainder, dual-use systems, could be dealt with accordingly.

For example, the canonical space-based weapon is a missile defense system designed to target long range missiles as they are launching. The Strategic Defense Initiative President Reagan initiated in 1983 and the Brilliant Pebbles program of the 1990s developed concepts and technologies for such a system, including space-based lasers and interceptors, though no systems were ever deployed. Because even an intercontinental-range ballistic missile has only 3–5 minutes of active launch time, these weapons would have to react quickly and propel the weapon toward their target within a few minutes. So the weapons themselves require powerful engines to get them out of orbit and to the right place at the right time. Such weapons would look unlike other satellites; their visual appearance and other clues, such as their orbital position, would indicate their function. While it is possible that they might be disguised or shrouded, this is unlikely to provide lasting secrecy. This is in part because such a system requires many satellites in a constellation in order to work. For example, a 2012 US National Academies of Science and Engineering study estimated that an "austere" space-based boost phase missile defense system to defend against one or two North Korean missiles would require 600 satellites (National Research Council 2012). In the same way, a space-based high powered

stated that debris-creating intentional satellite destruction should be banned (see: https://wien-io.diplo.de/iow-en/news/statement-debris/2208724).

[6]See, for example, the ABM Treaty (found at: https://2009-2017.state.gov/t/avc/trty/101888.htm) and the Intermediate Nuclear Forces Treaty (found at: https://2009-2017.state.gov/t/avc/trty/102360.htm).

laser designed to target earth objects or other space objects would require a very large source of power and large focusing optics. It would be difficult to mistake for anything else.

By the same token, while it is sometimes suggested that any satellite can be used as a weapon, because they are able to maneuver on orbit, this is simply not true in a practical sense. A modest amount of maneuvering capability is generally necessary for a satellite to keep in its assigned spot or to change orbits for a new mission, but maneuvering is slow and limited and not precise enough to intentionally get very close to another satellite. The equipment for rapid or large-scale maneuver, or the ability to get in close proximity to another satellite is specialized and would take up space and cost money. If a satellite has more capability than is required for its nominal mission, it may then be marked for further investigation to ensure that it is either not usable or being used as a weapon.

Which leads to a third category, satellites that could be used as weapons but might be dual purpose, the typical example being rendezvous and proximity operations satellites. Arms control can treat such systems separately, by agreements on how such satellites are operated, by requiring that the satellite's capabilities are sized for their stated purpose.

This suggests that should states decide it is important to negotiate limits on space-based weapons then there are viable paths to do so.

7.3.2.2 Verification

As discussed above, verification regimes can include a wide range of tools, from passive observations to more intrusive instruments such as on the ground or in space inspections. Regimes can be constructed during the negotiation of the agreement, tailored to the agreement and with a provision to revise and consult.

Many of the relevant space weapons and anti-satellite technologies are well-suited to being managed in a verification regime. For example, as mentioned above, the launching missile of a ground-based interceptor weapon creates a bright signal visible from space that cannot be disguised and a space-based monitoring system could be arranged to detect and assess all such launches.

Verification and monitoring will present a set of technical challenges, including those of detection and attribution. In some cases the impediments may be lack of political will rather than technical showstoppers, as increasing confidence may require increasingly intrusive verification methods. However, because weapons will need to apply force or energy at relatively long distances (with the exception of those on close-approaching satellites), the observable phenomena may be more visible the more lethal a technology is. This may make the most problematic technologies the simplest to detect. For example, ground-based lasers can be designed to interfere with an observation satellite's sensors. Such interference would cause the imaging satellite not to be able to perform its function over a limited time and geographical area, or at worst damage part of the sensor if the satellite had not included protections. These systems are fairly low-technology and do not require much power and so the presence of these satellite dazzling systems may remain undetected before use, but once they are used, their existence is made apparent and offender identified.

But ground (or space) based laser systems that could impart enough energy on a satellite's body or subsystems to inflict permanent disabling or damage would require orders of magnitude more power and sophistication than a dazzler would (Wright et al. 2005). Such a system would require a facility to house the laser and an energy system to produce the power necessary to operate it, and a large sophisticated tracking mirror with an adaptive optics system. For those reasons, the existence of such systems would be much more prone to detection during development and testing.

7.4 Getting Started

For the foreseeable future, the US and Russia and China will have difficult relationships, and they certainly do not currently agree on an approach to space arms control. The prospects of a comprehensive space arms control process leading to a successful result seem slim. However, simply starting the process anyway can have real benefits. A high-level discussion with real stakes will require states to get the perspectives and buy-in from key domestic stakeholders, to assemble technical and diplomatic experts for a negotiating team, and refine policy positions. While ad hoc processes in which one state makes unilateral declarations, or creates norms and tries to get its allies to agree can provide some incremental advantages, a process that requires a state to identify stakeholders and to get them to substantively engage may yield more substantial results.

An informed negotiation can increase the understanding of states' interests, clarify areas of agreement, and identify disagreements that require more attention. Absent such discussions, technology developments may be interpreted in the worst possible light. For example, the US Congress gets yearly intelligence assessments, and the most recent one frames China as preparing to attack US satellites early, in an escalatory strategy (U.S.-China Economic and Security Commission 2019). This view was based on a Chinese strategic document. A careful read of the document (in Chinese) indicates that a more accurate interpretation of the described strategy is nearly the opposite of this—that China sees these anti-satellite capabilities as securing deterrence against attacks on its own satellites (Kulacki 2019).

Perhaps most importantly, they can establish institutional relationships and communication channels. Because space has not been part of a major peer conflict, there's little basis for understanding adversary's perspective or putting their behavior in space into context in a crisis. A veteran diplomat, in comments written after taking part in a 2010 space "war game" simulated conflict described how this lack of experience would affect the management of a crisis if one or more militarily important U.S. satellites failed (Bloomfield Jr 2010):

> There is, in this future scenario, no credible basis for anyone around the president to attribute restraint to the adversary, no track record from which to interpret the actions by the adversary. There is no crisis management history: the president has no bilateral understandings or guidelines from past diplomatic discussions, and no operational protocols from previous incidents where space and cyber moves and counter-moves created precedents.

This experience and these relationships are important to interpreting events and to resolving disputes before they turn into a crisis, and to managing one once it begins.

There's nothing easy about such a process. Arms control meant to prevent an arms race, or to enhance stability and avoid crisis escalation, or to protect the space environment will prioritize different things. The three most important space users do not agree on next steps. Russia and China have a strong preference for working through the issue in a multilateral forum by negotiating a legally-binding and comprehensive agreement, while the United States has long expressed preference for a gradual establishment of norms instead. Additionally, the Trump administration has been overtly hostile to arms control, in two years initiating the US withdrawal from two treaties and a multilateral agreement.[7]

Given this, making progress will be challenging and will take sustained effort, but the impediments do not arise from a lack of viable ideas nor from unique particularities of the space environment that render it not amenable to arms control. Success will certainly require strong leadership, which looks unlikely from the three largest space powers, at least in the near term. A nascent effort by the "space middle powers"may break through this logjam of political will (Archinard 2016).

This chapter has outlined the risks that national security space operations can pose on earth and in space, which should provide strong motivation to pursue solutions. It has identified, as well, some ways that arms control can clearly help mitigate those risks by, for example, decreasing the threats satellites face by limiting the most dangerous technologies; heading off misperceptions that can be dangerous in a crisis by increasing communication and transparency; and avoiding military activities that might result in significant contamination of the space environment with debris.

Finding a way forward is critical for providing for a future in space and on earth that is more secure and more sustainable.

References

Archinard, N. (2016). *Space security and governance – The role of space middle power*. 10th United Nations Workshop on Space Law. https://www.unoosa.org/pdf/SLW2016/Panel2/4._Archinard_Presentation_NArchinard_UNOOSA_workshop_16.9.16.pdf

[7]The Trump administration announced the US withdrawal from the Joint Comprehensive Plan of Action May 8, 2018; the Intermediate Nuclear Forces agreement on February 2, 2019; and the Open Skies Treaty on May 21, 2020.

Beard, J. (2017). Soft Law's failure on the horizon: The international code of conduct for outer space activities. *University of Pennsylvania Journal of International Law, 38*(2), 335.
Bloomfield, A. L. P., Jr. (2010). National Security Fundamentals in the space and cyber domains. *High Frontier, 7*(1), 34–35.
Brooks, L. F. (2020, March 23). The end of arms control? *Daedalus, 149*(2), 84–100.
EEAS. (2015). *Security and sustainability in outer space*. https://eeas.europa.eu/headquarters/headquarters-homepage/8466/security-and-sustainability-outer-space_en
ESA. (2019, July 17). ESA's annual space environment report. https://www.sdo.esoc.esa.int/environment_report/Space_Environment_Report_latest.pdf
Grego, L. (2011). The AntiSatellite capability of the phased adaptive approach missile defense system. *Federation of American Scientists Public Interest Report*. https://fas.org/pubs/pir/2011winter/2011Winter-Anti-Satellite.pdf
Hansen, B., Starchville, T., & Hoots, F. (2015). First responders in space: The debris analysis response team. *Crosslink: The Aerospace Corporation Magazine of Advances in Aerospace Technology, 16*(1), 20. https://aerospace.org/sites/default/files/2019-04/Crosslink%20Fall%202015%20V16N1%20.pdf.
Harrison, T., Johnson K., Roberts, T., & Young, M. (2020, March 30). Space threat assessment 2020. *Centre for strategic and international studies*. https://www.csis.org/analysis/space-threat-assessment-2020
Hennigan, W.J. (2020, February 10). Exclusive: Strange Russian spacecraft shadowing U.S. Spy satellite, general says. *Time*. https://time.com/5779315/russian-spacecraft-spy-satellite-space-force/
Henry, C. (2019, April 9). India ASAT debris spotted above 2,200 kilometers, will remain a year or more in orbit. *Space News*. https://spacenews.com/india-asat-debris-spotted-above-2200-kilometers-will-last-a-year-or-more/
Kulacki, G. (2019). The US Congress needs facts, not Hyperbole, on China's Space Program. *The Diplomat*.https://thediplomat.com/2019/11/the-us-congress-needs-facts-not-hyperbole-on-chinas-space-program/
Ministere des Armees. (2020). Florence Parly unveils the French defense space strategy. https://www.defense.gouv.fr/english/actualites/articles/florence-parly-devoile-la-strategie-spatiale-francaise-de-defense
Ministry of External Affairs (India). (2019, March 27). Frequently asked questions on Mission Shakti, India's Anti-Satellite Missile test conducted on 27 March 2019. https://www.mea.gov.in/press-releases.htm?dtl/31179/Frequently_Asked_Questions_on_Mission_Shakti_Indias_AntiSatellite_Missile_test_conducted_on_27_March_2019
Morgan Stanley. (2019, July 2). Space: Investing in the Final Frontier. https://www.morganstanley.com/ideas/investing-in-space
Munich Security Report. (2020). https://securityconference.org/publikationen/munich-security-report-2020/
National Defense Authorization Act for Fiscal Year 2020 (2019). https://www.congress.gov/bill/116th-congress/senate-bill/1790
National Research Council. (2012). Making sense of ballistic missile defense: An assessment of concepts and Systems for U.S. In *Boost-phase missile defense in comparison to other alternatives*. Washington, DC: The National Academies Press. https://doi.org/10.17226/13189.
PTI. (2019, March 27). Narenda Modi announces success of Mission Shakti, India's anti-satellite missile capability. *The Hindu*. https://www.thehindu.com/news/national/narendra-modi-announces-success-of-mission-shakti-indias-anti-satellite-missile-capability/article26651731.ece
Schelling, T. C., & Halperin, M. H. (1961). *Strategy and arms control*. New York: Twentieth Century Fund.
Secure World Foundation. (2020, April). Global counterspace capabilities: An open source assessment. https://swfound.org/media/206970/swf_counterspace2020_electronic_final.pdf

Thompson, R. (2015 Fall). A Space Debris Primer. *Crosslink, The Aerospace Corporation magazine of advances in aerospace technology*. (pp. 4).
U.S.-China Economic and Security Commission. (2019). 2019 Annual Report to Congress. https://www.uscc.gov/annual-report/2019-annual-report-congress
UCS. (2010/2020). *Satellite database*. https://ucsusa.org/satellites
United States Department of Defense. (2020). *Defense space strategy summary*. https://media.defense.gov/2020/Jun/17/2002317391/-1/-1/1/2020_DEFENSE_SPACE_STRATEGY_SUMMARY.PDF p. 1 para 2
UNOOSA. (2010). Space debris mitigation guidelines of the committee on the peaceful uses of outer space. https://www.unoosa.org/pdf/publications/st_space_49E.pdf
UNOOSA. *Space supporting the sustainable development goals*. https://www.unoosa.org/oosa/en/ourwork/space4sdgs/index.html
US Department of Defense and Office of Director of National Intelligence. (2011). *National security space strategy unclassified summary*. https://archive.defense.gov/home/features/2011/0111_nsss/docs/NationalSecuritySpaceStrategyUnclassifiedSummary_Jan2011.pdf
Wright. (2007). Space Debris. *Physics Today*. https://physicstoday.scitation.org/doi/10.1063/1.2800252
Wright, Grego, Gronlund. (2005). The physics of space security. *American Academy of Arts and Sciences*. https://www.ucsusa.org/resources/physics-space-security

Chapter 8
Space Dispute Resolution Mechanisms

Stacey Henderson

Abstract There are very few rules for the resolution of disputes set out in international space law. Existing international space law contains only a fragmented, patchy and out-dated system for the resolution of disputes relating to space activities. This chapter outlines and evaluates the existing international dispute resolution mechanisms available to resolve space disputes. It also considers whether there is a need to establish a new *sui generis* international dispute settlement body to adjudicate disputes relating to space activities.

8.1 Introduction

When the international space law regime was created, the current and ever-increasing number and diversity of actors undertaking space activities was never imagined. Indeed, the Outer Space Treaty was negotiated and entered into force at a time when there existed only two space actors, both of whom were States. While the dramatic increase in space activities by both government and commercial actors offers great hope to humanity – for our species, for scientific endeavours, and for commercial purposes – they also carry with them the increased likelihood of disputes between participants in these ventures, as well as between participants and third parties.

This chapter suggests that the mechanisms for resolution of space disputes should be considered, reviewed, and updated before there is a proliferation of disputes arising in connection with space activities. Doing so would allow the design and integration of pro-active rather than reactive dispute settlement options. Indeed, commercial operators, and particularly insurers, may regard the creation of such dispute resolution mechanisms as a necessary pre-condition for investment in further space activities. This chapter first outlines and evaluates the existing international

S. Henderson (✉)
Adelaide Law School, The University of Adelaide, Adelaide, SA, Australia
e-mail: stacey.henderson@adelaide.edu.au

M. de Zwart, S. Henderson (eds.), *Commercial and Military Uses of Outer Space*, Issues in Space, https://doi.org/10.1007/978-981-15-8924-9_8

dispute resolution mechanisms available to resolve space disputes. It then considers whether there is a need to establish a new *sui generis* international dispute settlement body to adjudicate disputes relating to space activities. In advocating for the creation of a bespoke space dispute resolution mechanism, this chapter argues that the peaceful settlement of space disputes can be encouraged through the creation of a fit for purpose, specialised dispute resolution mechanism. The availability of such a dispute resolution mechanism can reduce the likelihood of escalation into conflict and hostilities, and foster an environment which is conducive to increased investment and innovation in the space sector for the benefit of all humanity.

8.2 Existing Dispute Resolution Mechanisms

The existing international space law treaties contain very few rules for the resolution of disputes relating to space activities (van Traa-Engelman 1990). Indeed, the limited rules that do exist have been described as a 'fragmented system for dispute resolution' (Hofman and Blount 2019). Further fragmentation of dispute resolution mechanisms, both in international law and between international and domestic law, is anticipated in the future as domestic regulators may seek to develop their own dispute resolution mechanisms and commercial dispute resolution providers are established. Hofman and Blount (2019) categorise the existing types of dispute resolution mechanisms as treaty-based, institutional, and judicial/quasi-judicial, with the majority of these mechanisms being available only for States.

There is no clear definition of what a dispute relating to space activities is; not every dispute related to a space activity will be a space dispute. There will be some disputes that arise in connection with a space activity that would fall well outside of the scope of international space law and which could be resolved utilising existing international or domestic dispute resolution mechanisms.[1] Space disputes could relate to collisions between space objects, disturbance of historic lunar sites, contested landing sites, resource extraction, and many other space activities.

8.2.1 Outer Space Treaty

The *Outer Space Treaty* is the foundation of international space law, regulating the conduct of activities in space. It was negotiated and drafted at the height of the Cold War, and at a time when only the United States and USSR had space capabilities.

[1] For example, a dispute arising in connection with the employment contract of an individual who is an operator employed by a commercial space launch provider would not be a space dispute, but rather an employment dispute and could be resolved through domestic dispute resolution mechanisms.

During the drafting stage of the Outer Space Treaty, the USSR (1966) proposed the following draft article X on the resolution of disputes:

> In the event of disputes arising in connexion with the application or interpretation of the Treaty, the States Parties concerned shall immediately consult together with a view to their settlement.

Ultimately, however, no article on dispute resolution was included in the final agreed version of the Outer Space Treaty as consensus on this proposed article could not be reached (van Traa-Engelman 1990).

Under Article VI of the Outer Space Treaty, 'States Parties to the Treaty shall bear international responsibility for national activities in outer space... whether such activities are carried on by governmental agencies or by non- governmental entities' (Outer Space Treaty 1967). Further, that the 'activities of non-governmental entities in outer space... shall require authorization and continuing supervision by the appropriate State Party to the Treaty' (Outer Space Treaty 1967). This may have the effect of making States the only parties with standing in any dispute relating to space activities.

The Outer Space Treaty contains no articles specifically providing for dispute resolution mechanisms. However, traditional international law dispute resolution mechanisms are implicitly incorporated in the Outer Space Treaty (1967) by Article III which provides:

> States Parties to the Treaty shall carry on activities in the exploration and use of outer space, including the Moon and other celestial bodies, in accordance with international law, including the Charter of the United Nations, in the interest of maintaining international peace and security and promoting international cooperation and understanding.

This suggests that 'the regulations and procedures for the settlement of disputes in international law will apply' (van Traa-Engelman 1990) to disputes relating to space activities. Of particular note here is the explicit incorporation of the UN Charter, which lists various means for the peaceful settlement of disputes in Chapter VI, including, in Article 33, 'negotiation, enquiry, mediation, conciliation, arbitration, judicial settlement, resort to regional agencies or arrangements, or other peaceful means of their own choice' (United Nations 1945). This would leave open the possibility of States Parties bringing a dispute relating to space activities before the International Court of Justice. Although, as a Court for States only, this dispute resolution mechanism would not be open to commercial or private space actors, unless they had previously exhausted all domestic remedies (Interhandel 1959) and were able to convince their State to commence a dispute in the International Court of Justice on their behalf. It could be extremely difficult to persuade a State to do so, particularly in circumstances where the State has no interest in the space activity that is the subject of the dispute. Additionally, the International Court of Justice does not appear to be a suitable forum to resolve space disputes which may be time critical and involve parties beyond the State and assets which are off-Earth. No space disputes have yet been referred to, heard or decided by the International Court of Justice.

The Outer Space Treaty contains a vague reference to dispute resolution in the context of harmful interference, although this seems focussed more on the prevention of a dispute rather than dispute resolution. Article IX urges States Parties to engage in 'appropriate international consultations' (Outer Space Treaty 1967) in situations of potential harmful interference with the space activities of another State Party. However, the actual procedure to be followed and what would amount to appropriate international consultation is unclear. Additionally, there is a scarcity of State practice to indicate when this consultation procedure should be followed, and, contrastingly, a growing amount of State practice indicating that States Parties to the Outer Space Treaty do not generally consider themselves obligated to engage in international consultations in relation to their space activities even when such activities may cause space debris.[2]

8.2.2 Liability Convention

The Liability Convention (1972) is the only one of the existing space law treaties to provide for a specific dispute resolution mechanism. Its preamble explicitly recognises:

> the need to elaborate effective international rules and procedures concerning liability for damage caused by space objects and to ensure, in particular, the prompt payment under the terms of this Convention of a full and equitable measure of compensation to victims of such damage.

The Liability Convention (1972) applies to a broad range of loss and damage, including 'loss of life, personal injury or other impairment of health; or loss of or damage to property of States or of persons, natural or juridical, or property of international intergovernmental organizations'. The treaty applies to loss and damage caused by the space object of a State Party, or its component parts, on Earth or in airspace, or in space, and also applies to loss and damage resulting from attempted launch of a space object.

The specific dispute resolution mechanism under the Liability Convention is facilitated through diplomatic channels, with claims for compensation for loss and damage to be presented by a claimant State to a launching State within a specified time frame of 1 year of the occurrence of the damage or the identification of the liable launching State. Article IX of the Liability Convention (1972) provides:

> A claim for compensation for damage shall be presented to a launching State through diplomatic channels. If a State does not maintain diplomatic relations with the launching State concerned, it may request another State to present its claim to that launching State or otherwise represent its interests under this Convention. It may also present its claim through

[2]The increasing State practice surrounding the conduct of anti-satellite (ASAT) tests by various States supports this view.

> the Secretary-General of the United Nations, provided the claimant State and the launching State are both Members of the United Nations.

Unlike many other international dispute resolution mechanisms, there is no requirement to have exhausted domestic remedies before commencing a claim under the Liability Convention procedure.

If the dispute cannot be resolved through diplomatic negotiations within 1 year from the commencement of the claim, the Liability Convention (1972) provides that either party to the dispute can request that a Claims Commission be established. The process for establishing the Claims Commission is set out in Articles XV to XVII. Once established, the role of the Claims Commission is to 'decide the merits of the claim for compensation and determine the amount of compensation payable, if any' (Liability Convention 1972).

There are several weaknesses in the dispute resolution mechanism provided for in the Liability Convention. First, total discretion to initiate an action lies with States Parties. Non-governmental entities cannot themselves commence a claim under the Liability Convention and are dependent on the willingness of a State Party to initiate a claim (Tronchetti, 2013). However, as van Traa-Engelman (1990) observes, this has been somewhat mitigated by the 'flexibility of the rule of nationality of claims . . . as an important step towards securing the interests of private parties through international law', with Article VIII providing that more than one State Party may be entitled to present a claim under the Liability Convention depending on the State's relationship with the natural or juridical person who has suffered loss or damage (Liability Convention 1972). Second, decisions of the Claims Commission will only be final and binding if the parties have so agreed (Liability Convention 1972). If the parties to the dispute have not agreed for the decision to be binding, it is merely a 'final and recommendatory award, which the parties shall consider in good faith' (Liability Convention 1972). Third, the types of disputes which will fall within its scope are quite narrow, being limited to claims for loss or damage as defined in Article I of the Liability Convention (1972). Finally, the dispute resolution mechanism provided for in the Liability Convention has never been used in practice (Tronchetti 2013), and, as such, its effectiveness as a dispute resolution mechanism has not yet been tested.[3]

[3]Disputes which might have been expected to activate the Liability Convention dispute resolution procedures have so far been resolved through other mechanisms. For example, the dispute relating to the 1978 crash of Cosmos 954 was resolved between the USSR and Canada through diplomatic channels; when parts of Skylab crashed in Australia in 1979, the United States' government was issued with a fine for littering; and the Liability Convention procedure was ruled out for the 2009 Iridium/Cosmos satellite collision due to the technical impossibility of proving fault.

8.2.3 *Moon Agreement*

For those States who are party to the Moon Agreement (1984),[4] there is the ability to request consultations as a means of dispute resolution. Article 15 of the Moon Agreement (1984) provides:

> 2. A State Party which has reason to believe that another State Party is not fulfilling the obligations incumbent upon it pursuant to this Agreement or that another State Party is interfering with the rights which the former State has under this Agreement may request consultations with that State Party. A State Party receiving such a request shall enter into such consultations without delay. Any other State Party which requests to do so shall be entitled to take part in the consultations. Each State Party participating in such consultations shall seek a mutually acceptable resolution of any controversy and shall bear in mind the rights and interests of all States Parties. The Secretary-General of the United Nations shall be informed of the results of the consultations and shall transmit the information received to all States Parties concerned.
>
> 3. If the consultations do not lead to a mutually acceptable settlement which has due regard for the rights and interests of all States Parties, the parties concerned shall take all measures to settle the dispute by other peaceful means of their choice appropriate to the circumstances and the nature of the dispute. If difficulties arise in connection with the opening of consultations or if consultations do not lead to a mutually acceptable settlement, any State Party may seek the assistance of the Secretary-General, without seeking the consent of any other State Party concerned, in order to resolve the controversy. A State Party which does not maintain diplomatic relations with another State Party concerned shall participate in such consultations, at its choice, either itself or through another State Party or the Secretary-General as intermediary.

Consultation as a means of dispute resolution under the Moon Agreement appears to include all situations where the activities of States Parties in the exploration and use of the Moon or other celestial bodies, on or below the surface, might generally interfere with the activities of other States Parties. This would include any disputes that arose between States Parties to the Moon Agreement in connection with resource extraction activities on the Moon. However, due to the very limited number of States Parties to the Moon Agreement, and the fact that none of the major space faring States are a party to the treaty, this mechanism for dispute resolution is unlikely to be of much practical use in resolving space disputes.

8.2.4 *Arbitration*

Following the Iridium/Cosmos satellite collision in 2009, it became clear that the existing dispute resolution mechanisms in international space law contained

[4]As at 30 June 2020, there are 18 States Parties to the Moon Agreement: Armenia, Australia, Austria, Belgium, Chile, Kazakhstan, Kuwait, Lebanon, Mexico, Morocco, Netherlands, Pakistan, Peru, Philippines, Saudi Arabia, Turkey, Uruguay, Venezuela. A further four States have signed, but never ratified, the Moon Agreement (France, Guatemala, India, and Romania).

significant shortcomings, with the 'technical impossibility of proving either party's fault' (Tronchetti 2013) ruling out the Liability Convention dispute resolution procedure for that dispute. While major institutions of international commercial arbitration, such as the United Nations Commission on Trade Law ('UNCITRAL') and the International Chamber of Commerce, have been utilised for space disputes, these fora are not particularly well suited as dispute resolution mechanisms for space disputes (Tronchetti 2013). In particular, space disputes involving issues relevant to national security or confidentiality over sensitive information may encounter difficulties under existing commercial arbitration procedures as these issues cannot be appropriately accommodated. There is also a lack of arbitrators with the necessary space-specific technical expertise to enable them to make an informed decision in relation to a space dispute (Tronchetti 2013).

Many existing intergovernmental agreements relating to space activities contain arbitration clauses, either for particular types of disputes or more generally.[5]The advantages of arbitration as a mechanism to resolve disputes relating to space activities have been noted to include the following:

- It is available to all parties, including individuals, corporations and States (Pocar 2012);
- The process is voluntary, operating on the consent of the parties, which may be expressed before a dispute arises by use of an arbitration clause. An arbitration clause may be included in any legal instrument including a treaty, an agreement between a State and industry, or in a commercial space contract between private entities (Pocar 2012);
- Decisions are final and binding, with no appeal mechanism and limited grounds for challenge (Pocar 2012);
- Arbitral awards are internationally recognised and enforceable for all 146 signatory States to the New York Convention (Tronchetti 2013);
- The procedure is flexible and able to be adapted to the needs of the parties (Pocar 2012);
- The parties can choose their own arbitrators, which can allow for the selection of arbitrators with specialised competence relevant to the specific space dispute (Tronchetti 2013); and,
- Arbitration hearings can be conducted in private, and awards may remain confidential, maintaining and safeguarding sensitive information and trade secrets (Pocar 2012).

In order to accommodate the unique aspects of space activities and the specific needs of actors in the space industry, the Administrative Council of the Permanent

[5]For example, the International Space Station Intergovernmental Agreement, International Telecommunications Satellite Organisation (INTELSAT), International Maritime Satellite Organisation (INMARSAT), European Space Agency (ESA), European Telecommunication Organisation (EUTELSAT), and the International Telecommunication Union (ITU) all provide for either compulsory or optional arbitration as a means of settling disputes either between Member States or with third parties.

Court of Arbitration adopted the Optional Rules for Arbitration of Disputes Relating to Outer Space Activities in December 2011 (Permanent Court of Arbitration 2012). The Optional Rules provide a means of voluntary and binding dispute resolution for space-related disputes tailored to 'the particularities of space-related disputes' (Tronchetti 2013). They were developed by a group of international experts and are based upon the arbitration rules of the Permanent Court of Arbitration and UNCITRAL modified to accommodate the unique characteristics of space activities.

The Optional Rules aim to provide the timely resolution of space disputes, which is particularly important for disputes involving time sensitive space activities, such as launch, orbital insertion, and re-entry (Tronchetti 2013). The scope of the Optional Rules is very broad; all actors involved in space activities, including States, intergovernmental organisations, nongovernmental organisations, corporations, and private entities, are able to use them to resolve their space dispute (Tronchetti 2013). The highly technical nature of space activities and the difficulties that this may pose in resolving space disputes is specifically accounted for in the Optional Rules. Article 27(4) of the Optional Rules states that the arbitral tribunal can request that the parties provide a 'non-technical document summarizing and explaining the background to any scientific, technical or other specialized information which the arbitral tribunal considers to be necessary to fully understand the dispute' (Permanent Court of Arbitration 2012). The confidentiality provisions of the Optional Rules are also uniquely suited to the space industry and space disputes. Article 17(8) provides that a confidentiality advisor can be appointed by the arbitral tribunal, either at the request of a party or on its own initiative, to report to the tribunal on specific issues on a confidential basis, without revealing the content of any confidential information either to the tribunal or the parties (Permanent Court of Arbitration 2012).

> The innovative character of the Rules is that, if parties choose to settle their dispute by arbitration and they agree to use the PCA as a forum, they may benefit from a set of procedural rules for arbitration specifically designed for outer space activities which ensures broad accessibility, a high level of expertise, and a binding conclusion of the dispute (Tronchetti 2013).

8.3 Draft Convention on the Settlement of Space Law Disputes

The International Law Association ('ILA') recognised the limitations in the existing dispute resolution mechanisms for disputes relating to space activities as long ago as 1978 (van Traa-Engelman 1990). Indeed, after extensive deliberations, the Space Law Committee of the ILA prepared a Draft Convention on the Settlement of Space Law Disputes (ILA 1998). The 1998 Draft Convention provided for the creation of an International Tribunal for Space Law, and promoted conciliation and arbitration as suitable dispute resolution mechanism for space disputes.

The 1998 Draft Convention would have applied to a broad range of space disputes, including those relating to all activities in space and all activities with effects in space, whether such activities were carried out by a State Party, by nationals of a State Party, or from the territory of a State Party (ILA 1998). The broad scope of the 1998 Draft Convention could be modified and constrained by States Parties, by excluding specific space activities, or by making a declaration to limit the scope of its application (ILA 1998). The 1998 Draft Convention would not apply when parties had agreed 'to submit their dispute to another procedure of peaceful settlement entailing a binding decision' (ILA 1998).

A series of dispute resolution procedures, both binding and non-binding, were proposed in the 1998 Draft Convention. The non-binding settlement procedures included an expeditious exchange of views, and conciliation (ILA 1998). The binding settlement procedures, which could only be utilised after the exhaustion of all available domestic remedies, offered States Parties a choice of forum for their dispute in Article 6 (ILA 1998), including:

(a) the International Tribunal for Space Law, if and when such a Tribunal has been established in accordance with section VI;
(b) the International Court of Justice;
(c) an arbitral tribunal constituted in accordance with section V.

The decision of any Court or Tribunal would have been binding on the parties in respect of the particular dispute, and would be final (ILA 1998).

Article 8 of the 1998 Draft Convention allowed a Court or Tribunal to select at least two scientific or technical experts, either at the request of a party or on its own initiative, to sit with the Court or Tribunal in any disputes involving scientific or technical matters (ILA 1998). Any such expert would not have had any voting rights, but would have been able to provide assistance to the Court or Tribunal on scientific or technical matters.

The creation of the proposed International Tribunal for Space Law was outlined in section VI of the Draft Convention (ILA 1998). It was proposed in Article 38 that there would be 21 independent members of the Tribunal, 'elected from among persons enjoying the highest reputation for fairness and integrity and of recognized competence in the field of space law' (ILA 1998). There was to be no more than two Tribunal members from the same State, and no fewer than three Tribunal members from each geographical group (ILA 1998) to ensure 'representation of the principal legal systems of the world and equitable geographical distribution' (ILA 1998).

The dispute settlement mechanisms provided for in the 1998 Draft Convention contained the possibility, with the exception of matters referred to the International Court of Justice, that there could be parties other than States Parties to the Convention involved in a space dispute. The arbitration section of the Draft Article explicitly provided that 'provisions of this section shall apply *mutatis mutandis* to any dispute involving entities other than State Parties' (ILA 1998). As van Traa-Engelman (1990) has noted in relation to an earlier draft of the Convention:

> the very possibility of presenting their case to an arbitral tribunal in order to receive a binding decision on a matter of international space law, can be seen as an important step towards a recognition of private enterprise in a domain, which, until now was purely restricted to states.

Ultimately, however, there was little support from States for the ILA proposal; the Convention never progressed beyond the Draft Convention stage and the proposed International Tribunal for Space Law was never created.

8.4 A *sui generis* Dispute Resolution Mechanism for Space Disputes

As the 'commercial sides of space activities are growing on an unprecedented scale' (Williams 2015), the possibility of disputes increases and the existing dispute resolution mechanisms threaten 'to become less effective' (Williams 2015). There is a need for a *sui generis* dispute resolution mechanism for space disputes 'tailored to the particularities of this unique area of economic activity' (Pocar 2012), and which has jurisdiction to adjudicate space disputes. This chapter will refer to this proposed international court or tribunal for space disputes as the 'International Space Court'. As with any international court or tribunal, a dedicated International Space Court would need to be created by consent through international agreement, either through the United Nations or by treaty. This appears unlikely in the current international climate where there is little appetite for new treaties. However, with the increase in space activities and actors, and the push for both sustained human presence in space and resource extraction, it is anticipated that there will be a need in the future for a dedicated International Space Court.

The International Space Court would operate in addition to existing international disputes resolution mechanisms, and would provide space actors, particularly commercial and individual actors, with a dedicated forum to resolve their space disputes. It would not deal with matters that would fall within the jurisdiction of the International Criminal Court. Any criminal matters that arose in the context of a space activity would fall within the jurisdiction of domestic Courts, either under international space law, the principle of active nationality, or otherwise, or may even fall within the jurisdiction of the International Criminal Court. The International Space Court would also not deal with matters that would fall within the jurisdiction of the International Court of Justice, or that the parties had chosen to take to arbitration under the Optional Rules (Permanent Court of Arbitration 2012).

In an effort to encourage State support and engagement with the International Space Court, the procedures adopted would need to include compulsory mediation and facilitated negotiation as a compulsory part of the dispute resolution process. In this way, the peaceful settlement of disputes can be encouraged while promoting the methods of alternative dispute resolution favoured by States and more likely to lead to outcomes that will allow space actors to continue to work together on space activities. Although the space sector is growing, and will continue to do so, there are a finite number of companies and individuals with the necessary skills and products required for successful space missions. As such, maintaining relationships must be an important part of any dispute resolution process for space disputes.

The International Space Court suggested would not be a permanent Court with a permanent Court room. What is proposed is an agile dispute resolution body (Williams 2015), able to respond to the needs of parties for quick access to dispute resolution processes regardless of where they may be located, whether on or off-Earth, and with an emphasis on alternative forms of dispute resolution such as mediation and facilitated negotiation. Space activities are time critical, and any dispute resolution process involving a space dispute must be designed and established with this in mind.

New technology could be used to manage space disputes, as well as conduct hearings and alternative dispute resolution processes. The ability to use a virtual Court will be increasingly important in the future for any disputes involving off-Earth human settlements, where it is simply not feasible for the parties to travel to the same physical location to have their dispute heard. In conducting dispute resolution processes using technological means, costs could be reduced as it would not be necessary to have a large permanent facility for the International Space Court; more modest accommodation for the registry and facilities for the Judges would suffice. This also raises the possibility of Judges and other staff of the International Space Court not needing to co-locate with the Court, facilitating recruitment of the most highly qualified candidates regardless of geographical location and ability to travel. Regional centres could be established, including in off-Earth human settlements, offering the necessary technologies to facilitate the virtual dispute resolution processes, including mediations, negotiations, and even hearings. This would not require that the parties to the dispute travel to the location of the regional facility for their dispute, the secure technology being used for the virtual Court could be coordinated from the regional facility with the parties being physically located elsewhere, or even in disparate locations. Keeping with the notion of an agile dispute resolution procedure, all Court processes for matters proceeding to determination should be conducted electronically. Court documents should be filed using secure portals, and witness evidence given via virtual means.

Given the complex technologies involved in space activities, and the unique elements of international space law, any judges of the International Space Court must be experts in international space law specifically, not international law more generally. Judges should be nominated by States Parties to the agreement establishing the International Space Court. Each State Party would be entitled to nominate one Judge for election by absolute majority vote by secret ballot of States Parties, and Judges should be appointed for a 5 year term. There should be an odd number of Judges to hear a dispute to avoid deadlocked decisions, and the bench for contested matters should contain no fewer than 7 Judges. While the number of Judges may seem large when compared to domestic courts and tribunals, this is commensurate with other international courts. In addition to determining space disputes, Judges would be expected to lead compulsory mediations and facilitated negotiations, which would be conducted by a single Judge in order to expedite the dispute resolution process and increase the availability of these mechanisms. As such, it is crucial that the Judges nominated by States possess the necessary skill set and expertise to successfully conduct alternative dispute resolution processes.

Space disputes involve complex issues of law and fact, and are likely to include complex technologies, some of which may be protected trade secrets or have national security implications. It is imperative that involvement in any dispute resolution process does not require the disclosure of trade secrets in an open forum nor undermine the ability of companies involved in the space industry to maintain their competitive advantage by protecting their trade secrets. While transparency is an important aspect of any dispute resolution procedure, this must be balanced against commercial needs. Accordingly, the International Space Court should allow parties the option of conducting closed or partially closed hearings, and the compulsory mediation and facilitated negotiation processes should be conducted on a confidential basis. The International Space Court should also have the power to instruct an independent, impartial technical advisor to disclose to the Judges and parties only that part of any secret technical information that is necessary for the resolution or determination of the dispute. The technical advisor should be required to enter into a non-disclosure agreement preventing the sharing or disclosure of the technical information before being entrusted with any confidential or secret information, and there should be serious contempt penalties imposed for breach.

The International Space Court should adopt a cost reduction approach, similar to the structure adopted by the International Criminal Court (Pejic 1998). By only sitting when a matter is before the court, ongoing operation and running costs can be minimised. It is anticipated that there will be high demand for alternative dispute resolution processes, and that this will form a major part of the work of the Judges. Equal access to dispute resolution mechanisms is important; it is critical that the costs associated with bringing a matter to the International Space Court, whether for mediation, facilitated negotiation or determination, is not beyond the reach of start-ups or individuals. As such, it is proposed that the International Space Court be funded by contributions from States Parties to the agreement establishing the Court, and that filing fees are kept modest.

The enforcement of any judgments of the International Space Court remains a challenge, particularly if assets or parties to a dispute are located off-Earth. Space disputes already have the potential to include objects that are beyond Earth's atmosphere. Unless there is some way of enforcing a judgment off-Earth, there seems little point in bringing a claim against an individual in any future off-Earth human settlement. To address this, the recognition and enforcement of judgments of the International Space Court, as well as judgments of international courts and tribunals generally, should be included in the governance arrangements for any future off-Earth human settlements.

8.5 Conclusion

Space activities are high cost and high risk. They typically involve complex technologies and multiple actors ranging from States, to private companies and individuals from multiple States. The diversity of actors involved in space activities will only increase in the future, and it is imagined that, in the not too distant future, the potential parties to a space dispute could commonly include parties based on different planets and objects manufactured in space.

The development of special dispute resolution mechanisms for space disputes before they become an international (or intergalactic) crisis, offers the opportunity for pro-active, rather than reactive, dispute settlement options. Establishing 'an effective mechanism for dispute settlement in international space law could positively influence the general subject matter of peaceful settlement of international disputes by setting an example for international relations in the increasingly dependant world of modern times' (van Traa-Engelman 1990). Indeed, some States may seek to establish themselves as the forum of choice for dispute relating to space activities by developing bespoke dispute resolution mechanisms for space disputes that are tailored to the unique nature of space activities.

This chapter has evaluated the existing disputes resolution mechanisms for space disputes, and, with the exception of arbitration under the Optional Rules, found them lacking. The chapter has called for the establishment of a dedicated International Space Court to encourage the peaceful settlement of disputes and reduce the likelihood of escalation of disputes into conflict and hostilities. By encouraging early resolution of space disputes through alternative dispute resolution processes, the International Space Court could foster an operational environment of certainty and stability, conducive to increased investment and innovation in space activities for the benefit of all humanity.

References

Agreement Governing the Activities of States on the Moon and Other Celestial Bodies. (1984). ('Moon Agreement').

Convention on International Liability for Damage Caused by Space Objects. (1972). ('Liability Convention').

Hofman, M., & Blount, P. J. (2019). *Space law disputes*. Max Planck Encyclopedias of International Law. http://opil.ouplaw.com. Accessed 29 June 2020.

Interhandel (Switzerland v United States of America). (1959) ICJ Rep 6.

International Law Association. (1998). Final draft of the Revised Convention on the Settlement of Disputes Related to Space Activities. *Report of the 68th Conference, Taipei, Taiwan*. pp. 249–267.

Treaty on Principles Governing the Activities of States in the Exploration and Use of Outer Space, including the Moon and Other Celestial Bodies. (1967). ('Outer Space Treaty')

Pejic, J. (1998). Creating a permanent international criminal court: The obstacles to Independence and effectiveness. *Columbia Human Rights Law Review, 29*(2), 291–354.

Permanent Court of Arbitration. (2012). *Optional rules for arbitration of disputes relating to outer space activities*. A/AC.105/C.2/2012/CRP.17.

Pocar, F. (2012). An introduction to the PCA's optional rules for arbitration of disputes relating to outer space activities. *Journal of Space Law, 38*(1), 171–210.

Tronchetti, F. (2013). The PCA rules for dispute settlement in outer space: A significant step forward. *Space Policy, 29*, 181–189.

United Nations. (1945). Charter of the United Nations.

USSR. (1966). *Draft Treaty on Principles Governing the Activities of States in the Exploration in the Use of Outer Space, the Moon and Other Celestial Bodies*. U.N. Doc. A/6352.

van Traa-Engelman, H. (1990). Settlement of space law disputes. *Leiden Journal of International Law, 3*(3), 139–156.

Williams, M. (2015). Dispute resolution regarding space activities. In F. von der Dunk & F. Tronchetti (Eds.), *Handbook of space law* (pp. 995–1046). Cheltenham/Northampton: Edward Elgar.

Chapter 9
Connecting Diverse Data to Address Emerging Space Sector Challenges

Samantha Le May

Abstract Innovative new technologies and business models are continuing to reduce the cost of launch and manufacture of spacecraft, paving the way for increased participation in the space industry by corporations, start-ups, and emergent space economies. As Earth-orbital environments become increasingly populated, Space Traffic Management capabilities will be key to coordinating international space activities. To preserve the long-term sustainability of the outer space environment, new approaches that can leverage diverse data from multiple information sources to generate new insights are needed to inform decision making in the space sector. This chapter describes the current population of space objects (including satellites and space debris), details existing approaches to Space Traffic Management, identifies emerging challenges for the space sector and proposes the use of graph databases as an analytic tool to inform decision making in the space domain.

9.1 Space Objects in Earth Orbit

On Friday, 4 October 1957, the USSR launched Sputnik, Earth's first artificial satellite. This milestone achievement sparked the beginning of the Space Race – a competition between Cold War rivals to land on the moon, with victory claimed by the successful Apollo 11 moon landing in July of 1969. Since then, developments in satellite technology have progressed expansively to form today's vast network of operational satellites. According to ESA's *Space Environment Statistics*, this has culminated in around 5,950 rocket launches placing around 9,980 satellites into Earth orbit. Of these, around 5,650 satellites remain in space and about 2,800 continue to function. Space debris make up the majority of catalogued space objects with about 26,550 debris objects regularly tracked by Space Surveillance Networks.

S. Le May (✉)
SPACE Research Centre, RMIT University, Melbourne, VIC, Australia
e-mail: samantha.lemay@rmit.edu.au

M. de Zwart, S. Henderson (eds.), *Commercial and Military Uses of Outer Space*,
Issues in Space, https://doi.org/10.1007/978-981-15-8924-9_9

Estimates indicate more than 550 events have taken place causing the fragmentation of satellites by explosions, collisions or anomalies. The number of these fragments, most of which are below the threshold of tracking capabilities, are estimated at more than 128 million objects (ESA 2020). The following sections provide an overview of functional satellites and debris objects in orbit.

9.1.1 Functional Satellites and their Applications in Military, Government, Civil and Commercial Sectors

Satellite technology has a diverse range of general applications with future developments certainly providing even greater potential. In a practical sense, the average civilian benefits greatly from global navigation satellite systems, communications, entertainment broadcasting and weather forecasting services enabled by satellite technology. Less apparent, however, are the significant contributions satellite services provide to the security and well-being of our global population.

The Union of Concerned Scientists (UCS) have prepared and maintained an open-source database of active satellites since 2005 to provide detail of operational satellites currently in orbit as a tool for research. Their database includes detail about the satellite mission type and assigns the affiliation of its users as either Civil, Commercial, Military or Government, or some combination of these. Figure 9.1 summarises combined data from the UCS database, the Commercial Space Operations Centre (CSpOC) public space object catalogue, and Jonathan's Space Report, to show the total number of Satellites launched each year by each sector. In 2017, a significant increase in overall activity is observed with close to 400 satellites launched compared to 200 or fewer satellites launched per year historically. The

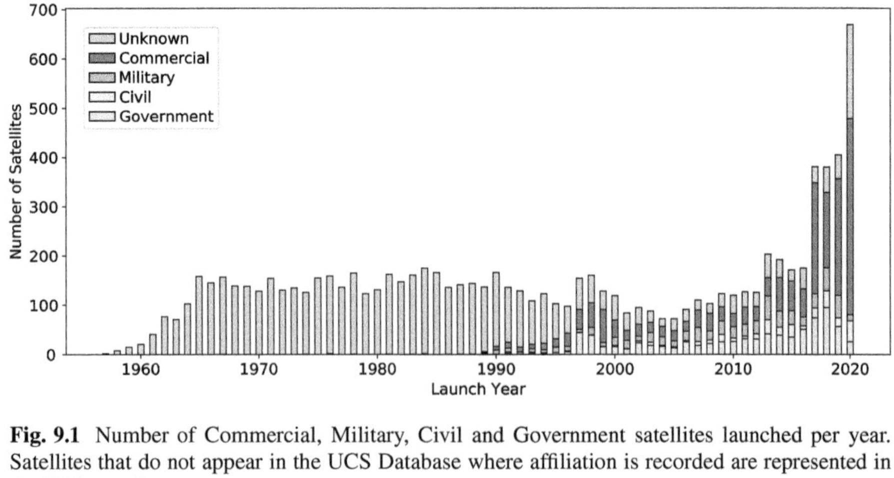

Fig. 9.1 Number of Commercial, Military, Civil and Government satellites launched per year. Satellites that do not appear in the UCS Database where affiliation is recorded are represented in the 'Unknown' category

substantial growth of the Commercial space sector compared to others is also apparent over the previous decade. As at 30 June 2020, an unprecedented 650 satellites have already been launched with around 400 of these owned commercially.

To more closely examine the composition of the active satellite network, Tables 9.1 and 9.2 consider only the active satellites listed in version 1 April 2020 of the UCS Satellite database. Of the 2,666 included in the dataset, Commercial users represent the largest group with ownership of approximately 54% of the entire satellite network (Table 9.1). Government and Military follow with ownership of around 22% and 18% respectively. Civil ownership, consisting largely of academic affiliations and amateur users, represents the smallest group at around 5% or 136 satellites.

318 satellites in the dataset are affiliated with more than one user group, with Government/Commercial being the most common combined user group with 116 satellites total. Satellites listed in the UCS database as having both Government and Commercial users include communication satellite networks (USA's Iridium constellation, the ChinaSat communication network, and Russia's Gonets system) and Earth Observation satellites (South Korea's Kompsat, and Russia's Resurs-P series satellites operated by Roscosmos). The Military/Commercial user group of 79 satellites includes the USA's Navstar GPS network, and Russia's "GLObal NAvigation

Table 9.1 Users of active satellites

	Military	Government	Civil	Commercial	Total
Military	–	5	–	2	8
Government	55	–	2	1	58
Civil	6	43	–	11	60
Commercial	77	115	–	–	192
None	339	436	133	1440	2348
Total	477	599	136	1454	2666

Table 9.2 Users and purpose of active satellites

	Military	Government	Civil	Commercial	Total
Communications	129	173	18	888	1208
Earth observation	167	260	23	434	884
Tech development	55	82	71	92	300
Navigation	110	12	–	26	148
Science	3	63	17	2	85
Earth science	5	6	4	3	18
Tech demo	1	2	1	5	9
Space observation	7	1	–	–	8
Surveillance	–	–	–	3	3
Education	–	–	2	–	2
Mission extension tech	–	–	–	1	1
Total	477	599	136	1454	2666

Satellite System" (GLONASS). Combined Military/Government owned satellites are also common with this group comprised primarily of 49 Beidou Navigation satellites operated by the Chinese Defense Ministry. The Italian COSMO-Skymed constellation of Earth Observation satellites are also in this category, intended for both Military and Civil use.

Data from the UCS Satellite Database also includes information pertaining to the purpose of a satellite mission. Table 9.2 indicates that the primary purpose for 1,208 active satellites is for communication services, followed by 884 satellites providing services in Earth Observation, 300 satellites used as a platform for technology development and demonstration, and 148 satellites used for navigation services. The remaining 274 active satellites are employed in other areas including Science missions.

Communications is clearly the dominant application for satellites owned and operated by Commercial users, which include services such as telecommunications and messaging, broadcasting digital audio and video, and high-speed internet services. The largest contributors to the active satellite population in the Communications category are 358 Starlink constellation satellites which form part of the initial deployment of SpaceX's global broadband constellation. Global broadband satellite systems aim to provide high-speed and high-capacity internet service with global coverage, surpassing the limited coverage achieved by terrestrial broadband internet networks (ITU 2019). Improved internet and communications technologies form part of the UN's Sustainable Development Goals (SDGs) (United Nations General Assembly 2015). The ITU (2019) found that just 20% of people in Least Developed Countries currently have access to broadband internet, while over half of the total global population (51.2%), irrespective of development status, are connected. Their report identifies satellite communications as a critical component to provide communications infrastructure, in particular to connect rural and remote communities to internet services (ITU 2019).

Earth Observation satellites are common in both Military and Government user groups, which is understandable given the types of services provided by satellites operating with Earth Observation instrumentation. The monitoring, prevention and mitigation of events such as conflicts, acts of terror, earthquakes, drought, fires and severe storms, require a large volume of high-quality and detailed information. Technological developments in satellite technology have continued to improve the spatial, radiometric and temporal resolution of data provided by Earth-observing satellite networks (Jasani et al. 2009), with those improvements influencing the ability of Governments and Military to take actions to prevent or mitigate the risks posed to the security of affected populations at times of crisis.

In order to secure the wellbeing of a population, or to provide aid in the aftermath of disaster, the presence of a population must be established. In many low income countries, and in particular countries that have experienced prolonged conflict, population displacement or migration, population data can be sparse or even absent (Jasani et al. 2009). Satellite Earth Observation and geo-spatial techniques can be applied to identify geographical features such as buildings and transport infrastructure as an indication of human presence, to provide moderately accurate and current estimations of populations; as demonstrated by Harvey (2002).

Security of essential resources including food and water is another area that benefits from Earth Observation satellite technologies, which can be used to estimate population growth and forecast the future demand and availability for essential resources (Jasani et al. 2009). The information provided by high-resolution satellite imagery is critical not only where information is sparse, but also where events trigger dramatic changes to the landscape. In the aftermath of the 2010 Haiti earthquake, the collapse of the natural and built environment cantered around the city of Port-au-Prince posed significant challenges for emergency relief efforts. The timely accessibility to remote sensing data and satellite imagery is crucial to inform situation assessment, identify emergency focal points, and assess damage to infrastructure in order to resolve logistic challenges, as was in the case of the earthquakes in Haiti, 2010 (Pham et al. 2014).

Hurricane Sandy made landfall in October of 2012. The complex Category 3 storm system wreaked havoc across most of the eastern coast of the United States, leaving 233 fatalities in its path (Diakakis et al. 2015). An investigation by Mcnally et al. (2014) aimed to determine the importance of satellite data in successfully forecasting a sudden "left hook" taken by the path of the storm toward the coast of New Jersey. Their studies showed that without the contribution of observations from polar-orbiting satellites, the 4–5 day forecast would predict the storm following a path offshore in the Atlantic and making landfall on the coast of Maine 24 h later than the true event impacting the coast of New Jersey (Mcnally et al. 2014).

The above examples demonstrate only a small fraction of the invaluable services provided to society by satellite technology. Research continues to refine the capabilities of the world's earth-orbiting infrastructure, which in-turn will improve the quality and diversity of services offered by Commercial, Civil, Government and Military satellite owners. As space-based capabilities increase, so too will society's reliance on these new technologies in daily life – and as tools for the protection and aid of the population in the event of crisis.

9.1.2 Space Debris

In 1994 the United Nations subcommittee on the Peaceful Uses of Outer Space (UN COPUOS) considered the subject of characterizing the space debris environment as a priority. This led to the publication of UN COPUOS's first technical report on space debris in 1999 (Scientific and Technical Subcommittee 1999). Over time, the space environment has become increasingly polluted by orbital debris that threaten the safety of our vast network of space infrastructure. Sources of space debris include abandoned rocket bodies, decommisioned satellites, and fragments of spacecraft, as described in the following sections.

9.1.2.1 Rocket Bodies

There are a wide variety of launch vehicles used to place satellites into orbit which are either solid or liquid fueled and use one or more rocket stages to thrust the spacecraft to the required altitude (AESB 1995). For a Low Earth Orbit (LEO) satellite, typically only one rocket stage will remain in orbit at a relatively low altitude. In contrast, the launch of a satellite to Geostationary (GEO) orbit (35,000 km) may require three or more separate rocket stages, which are discarded at different orbital altitudes along the flightpath.

There are approximately 800 abandoned rocket stages currently in LEO ranging in size from 0.5 to 12 m in length, 0.5 to 4 m in diameter, and 30 to 9000 kg in dry mass (Anselmo and Pardini 2016). The orbits of the largest stages left at lower altitudes will typically decay and re-enter relatively quickly, but the stages used to transfer spacecraft into higher orbits can remain in orbit for very long periods of time (AESB 1995).

9.1.2.2 Decommissioned Satellites

As of 1 April 2020, the CSpOC catalogue recorded a total of 5,645 satellites in orbit (NASA Orbital Debris Program Office 2020). The 1 April 2020 version of the UCS Satellite Database reports that 2,666 of these satellites are operational. According to the IADC's definition of space debris objects, this classifies the 2,979 satellites that are no longer in operation as part of the space debris population. One of the largest of these inoperable payloads is Envisat, a remote sensing satellite operated by ESA, whose mission was concluded in May 2012 after attempts to re-establish communications with the satellite that were lost a month prior (ESA 2012). Approximately the size of a school bus, and with an initial orbit of around 700 km, Envisat is now destined to remain in LEO for an estimated 150 years before re-entering into Earth's atmosphere (Rosanelli 2014).

9.1.2.3 Fragmentation Debris

Fragmentation results from either surface deterioration or breakup events which may be caused by internal explosions due to malfunction, accidental collision with another object in orbit, or even intentional destruction (AESB, 1995). The process of fragmentation by collision has been identified as the primary contributor to the evolution and stability of the orbital debris environment (Klinkrad 2006). Intentional collision (ASAT), accidental collision or accidental explosion (due to propulsion failure) feature as the primary cause of the top fragmentation events (ranked by number of total fragments) in ESA's DISCOSWeb database (Table 9.3).

ASAT Unintentional collisions are not the sole contributor to the orbital debris populations; intentional destruction of satellites through anti-satellite (ASAT) weapon

Table 9.3 Top 5 fragmentation events by total number of fragments from ESA'S DISCOSWeb database (ESA 2019)

Name	Intl. designator	Breakup epoch	Cause	Total fragments	Fragments on orbit
Fengyun 1C	1999-025A	11 January 2007	ASAT	3431	2809
Cosmos-2251	1993-036A	10 February 2009	Collision	1667	1070
HAPS	1994-029B	3 June 1996	Propulsion	753	
Iridium 33	1997-051C	10 February 2009	Collision	627	326
Cosmos-2421	2006-026A	14 March 2008	Unknown	509	

testing is also a significant factor. ASAT testing has great potential to contribute detrimentally to the sustainability of near-Earth space, as was demonstrated by the Chinese FengYun 1C ASAT test which occurred in January 2007. This event resulted in more than 2000 trackable debris fragments (Newman et al. 2010), which is ten times the average observed increase of approximately 250 tracked objects per year (NASA Orbital Debris Program Office, 2007). This single event significantly increased the probability of collisions between debris objects and spacecraft (Bradley and Wein 2009), and has therefore been widely condemned amongst space experts (Bradley and Wein 2009; Durrieu and Nelson 2013; Pardini and Anselmo 2009). Today, 2,809 fragments from this single event remain on orbit Table 9.3, making it the single highest contributor to the Space Debris population.

Collision Fragments Fragmentation debris caused by breakup events have the potential to contribute a large volume of debris across a range of altitudes at relatively high velocities. The size, mass and velocity of impact are the parameters which determine the quantity of debris, size and velocity distribution of fragments. At high speeds, a collision between two large derelict satellites could have an energy release comparable to an atomic explosion, sending thousands of debris objects into orbits which will disperse over time. The challenge of predicting and preventing collisions in orbit is demonstrated by the accidental collision between Cosmos-2251 and Iridium 33 on 10 February 2009 (Pardini and Anselmo 2011). Both satellites were tracked by the former US Joint Space Operations Centre (now CSpOC) and the predicted miss distance using data available at the time of collision was reportedly 698 m, making the probability of collision extremely remote (Pardini and Anselmo 2011). Even a post-event analysis using additional observation data returned a miss distance of 223 m, and a negligible collision probability (Newman et al. 2009). More than a decade later, 1,396 space debris fragments originating from this event remain on orbit (Table 9.3). The Iridium-Cosmos collision exemplifies the large uncertainties still associated with the tracking and orbit prediction of objects in space, discussed further in Sect. 9.2.2.

Explosion Fragments On-orbit explosions of spacecraft and rocket bodies are a common cause for break-ups. An accidental explosion can result from malfunctions associated with the spacecraft's power or propulsion systems either during or after mission operations. Energy sources stored on board spacecraft in the form of residual propellants, compressed fluids, batteries, high-pressure vessels, flywheels and momentum wheels (IADC 2007), provide the fuel for an explosion event that can generate hundreds of debris fragments (Pelton 2013). The explosion of a Pegasus Hydrazine Auxiliary Propulsion System (HAPS) on 3 June 1996, which was not passivated at the end of its mission (Matney and Settecerri 1997), generated 753 fragments which have since re-entered Earth's atmosphere according to ESA's DISCOSWeb database 9.3.

Other sources Large quantities of aluminium are used in solid rocket fuel which, when burned, form large numbers of micron-sized aluminium oxide particles and a small number of 1 cm and greater lumps. These byproducts are known as solid rocket motor (SRM) slag which are added to the debris environment when SRMs boost satellites into higher orbits (AESB 1995). Frozen droplets of sodium-potassium (NaK) coolant are another source of debris that originate from the 16 nuclear-powered radar ocean reconnaissance satellites (RORSATs) launched by the former Soviet Union in the 1980s (Wiedemann et al. 2005). A reactor core ejection system unique to these spacecraft release the NaK coolant, forming frozen droplets up to a diameter of 5 cm concentrated between 900 and 950 km in LEO (Wiedemann et al. 2005).

9.2 Current Approach to Space Traffic Management

On the ground, road safety is achieved through a holistic and integrated management system encompassing the dynamic interactions between safe operating speeds, vehicles, road infrastructure and road user behaviour (World Health Organization WHO). Road safety strategies are implemented through effective policy tools including a strong emphasis on the enforcement of traffic laws through penalties, investment in road safety infrastructure, voluntary standards and guidelines, driver competency (including skills, expertise, capability and professional practice), education, innovation and research (Hughes et al. 2016). Broadly speaking, space lacks an enforceable "road safety strategy" and equivalent infrastructure such as highways, intersections and stop signs. Unlike cars on the road, satellites travel at high velocities without brakes, and are operated remotely from Earth without drivers in the front seat responding to hazards (such as collisions with debris and other spacecraft) in real time. Satellite operators instead rely on the concept of Space Traffic Management (STM), including Space Situational Awareness (SSA) capability for on orbit safety and space debris mitigation principles to sustain a safe environment for operations. The following sections provide an overview of current approaches to STM, SSA and space debris mitigation.

9.2.1 Space Traffic Management

The concept of STM evolved in discussions through the American Institute of Aeronautics and Astronautics (Antoni et al. 2020), eventually being defined by the 2006 IAA Cosmic Study on Space Traffic Management as "the set of technical and regulatory provisions for promoting safe access into outer space, operations in outer space and return from outer space to Earth free from physical and radio-frequency interference". More recently, the National Space Traffic Management Policy from the United States (2019) defined STM as "the planning, coordination, and on-orbit synchronization of activities to enhance the safety, stability and sustainability of operations in the space environment". While the definition and scope of STM is an ongoing discussion, the notion of "safety, stability and sustainability" are consistent in the views expressed recently by authors (Skinner 2019; Stilwell et al. 2020; Antoni et al. 2020) alongside advocacy for the establishment of standards of behaviour and best practices for space activities.

The concept of STM is inherently broad because it intersects both technical and governance aspects of operating in the complex and dynamic environment of Outer Space. While there is currently no international agreement on a mechanism for STM, society's growing dependence on space infrastructure has emphasised the need to assure safe space operations. The Guidelines for the Long-term Sustainability of Outer Space Activities put forth by the United Nations Committee on the Peaceful Uses of Outer Space (UN COPUOS) and adopted in June 2019, provides an international framework for STM through non-binding guidelines spanning technical and governance aspects to include: the safety of space operations; policy and regulatory frameworks for space activities; international cooperation, capacity-building and awareness; Scientific and technical research and development. The following sections summarise the role of SSA as an operational element of STM for the safety of space operations, and existing policy and regulation pertaining to Space Debris Mitigation.

9.2.2 Space Situational Awareness

'Situational Awareness' is a concept originating from dynamic fields such as aviation, disease control, and military operations which require timely knowledge about what is happening within a given space and time for the purpose of forecasting future events to mitigate threats and strategise (Kaiser 2015). In the space domain, SSA is arguably the most critical aspect for the safety of space operations, referring to the detection, characterisation and knowledge of space objects and the environment within which they operate. SSA activities include the maintenance of space object catalogues to track and disseminate information on the location and trajectories of objects in orbit, monitoring of space weather and environmental factors that can impact the trajectories or operation of those objects, and the screening of trajectories of relevant space objects for potential collisions (conjunction assessment). In a military context (also known as Space Domain Awareness), additional SSA activities can include intent and capability assessment, manoeuvre detection and ground track analysis. Historically a military function, today there are a growing number of commercial SSA service providers contributing to global efforts to catalogue space objects and provide conjunction assessments to prevent collisions in orbit.

One important factor in the provision of SSA services is the accuracy of shared information, the uncertainty associated with the location of objects in orbit, and the opportunity for various methods and models to produce different results. The predicted location of an object at a specified time is complex due to the large number of variables influencing the motion of an object in orbit including (but not limited to) its velocity, shape, mass, the density of the atmosphere and solar activity, all of which change over time. Assumptions are required to account for many of these variables – for example atmospheric density and solar activity are forecast by models with inherent uncertainty, and the shape of objects are often assumed to be a sphere. Further challenges arise when we consider that some spacecraft have the ability to manoeuvre, and that current prediction models do not account for potential routine behaviours of individual spacecraft or constellations.

When screening for a conjunction between two objects, the accuracy of model inputs will become a factor in the interpretation of the conjunction assessment that ultimately informs the decision of an operator to make a collision avoidance manoeuvre (assuming the spacecraft is equipped with manoeuvring capability). The challenges associated with conjunction assessment are highlighted by Nicolls et al. (2017) where the authors acknowledge that high-risk conjunctions are inevitable. They note that orbits derived using SSA data from the CSpOC catalogue are not accurate enough for collision avoidance, but were able to reduce uncertainties and increase confidence in manoeuvre-related decisions by including SSA data from other sources within their conjunction assessment workflow. Despite this improvement, uncertainties associated with the object trajectories in their study were still on the order of 10 to 100 s of metres, which reflects the current state of the art in tracking capabilities

The need for improved SSA capabilities is reflected in the US National Space Traffic Management Policy. The directive acknowledges that SSA inputs such as observational data, models and methods are critical to achieve this, but also highlights the need to develop standards and best practices including conjunction assessment protocols. While SSA is often viewed as a technical problem, it is evident that future success in this area will require effective policy and regulatory frameworks that enable SSA data sharing, standard protocols, and international cooperation in a globally inclusive STM regime.

9.2.3 *Space Debris Mitigation and Remediation*

Research predicting a rapid non-linear increase in LEO objects (Johnson et al. 2001; Reynolds et al. 1999; Walker et al. 2001, 2000) became motivation for the development of the first version of the Inter-Agency Space Debris Coordination Committee's (IADC) Debris Mitigation Guidelines in 2001. Recognising the benefit of a set of high-level qualitative guidelines to the global space community, the UN COPUOS developed international recommendations based on the technical content of the IADC's Debris Mitigation Guidelines (2007) in conjunction with existing UN treaties and principles on outer space (UN COPUOS. 2010). There now exists a number of different policy documents, guidelines and recommendations published at an international, national, and organisational level which have helped the space-faring community to mitigate the creation of new debris (Pace 2015). The following sections provide an overview of the space debris mitigation principles advocated by key organisations from an international perspective.

9.2.3.1 Principals of Space Debris Mitigation

There are four core principals of space debris mitigation that are common to all significant international guidelines (EU 2004; IADC 2007; UN COPUOS. 2010):

1. Limitation of debris released during normal operations
2. Minimisation of the potential for on-orbit break-ups
3. Post-mission disposal
4. Prevention of on-orbit collisions

These have been derived from the first and second versions of the IADC's Debris Mitigation Guidelines (2001 and 2004) and put forth by both the European Union's (EU) European Code of Conduct for Space Debris Mitigation (2004) and UN COPUOS's Space Debris Mitigation Guidelines (2010) as standard best practice for space operations undertaken by member states of the EU and UN respectively. Although guidelines are not legally binding, member states are encouraged to use these principles as a foundation for the development and implementation of national regulations. Operators of existing spacecraft in orbit are encouraged to apply guidelines to the greatest extent possible, while new space mission proposals should be guided by the standard principles of mitigation in their design, planning and execution (UN, 2016).

Limitation of debris released during normal operations Spacecraft and rocket bodies should be designed in such a way that limits the amount of debris released during normal operations. Any unavoidable release of debris should be minimised in number, cross-sectional area and orbital lifetime. In cases where mission-related debris objects will be necessary to operations, operators should ensure that any long-term effect on the orbital debris environment is assessed, and that the potential hazard to another spacecraft in orbit is acceptably low (EU 2004; IADC 2007; UN COPUOS. 2010).

Minimisation of the potential for on-orbit break-ups Minimising the potential for on-orbit break-ups is a necessary consideration both during operations and post-mission. At the end of mission, accidental break-ups resulting from stored energy can be avoided by passivisation, which requires the depletion of stored energy sources including residual propellants, compressed fluids, and electrical storage devices (UN COPUOS. 2010). During operations, accidental break-ups can be the result of malfunctions such as catastrophic failures of power and propulsion systems. It is therefore recommended that a spacecraft is monitored for malfunctions throughout the duration of their mission. Procedures should be in place for the passivisation and disposal of spacecraft, and executed when a malfunction is detected (IADC 2007). The intentional destruction of spacecraft in orbit (for example by ASAT testing) should also be avoided (EU 2004; IADC 2007; UN COPUOS. 2010). The IADC and UN COPUOUS guidelines advocate that intentional break-ups, if necessary, should be conducted at sufficiently low altitudes to limit the orbital lifetime of fragments.

Post-mission disposal After their end of mission, both the IADC and EU recommend the presence of spacecraft and rocket bodies in LEO should be limited to 25 years (EU 2004; IADC 2007). Space operators should undertake a thorough analysis to ensure that atmospheric drag will limit the orbital lifetime to achieve re-entry within the recommended 25 years. The preference, however, is for spacecrafts terminating in orbits that pass through the LEO region to be de-orbited in a controlled fashion (EU 2004; IADC 2007; UN COPUOS. 2010). The UN COPUOUS's guidelines do not reference a 25 year limitation achieved by natural orbit decay, instead recommending de-orbit manoeuvres or transferring the spacecraft to an uncongested region where their long-term presence will not have impact on the sustainability of the space environment (UN COPUOS. 2010).

Prevention of on-orbit collisions Spacecraft operators should estimate the probability of accidental collision with known objects during their system's launch phase and orbital lifetime. Hence, consideration should be taken to ensure the mission's specifications (e.g. orbital regime) will limit the likelihood of collision, and the spacecraft's physical design will limit the consequences of collision with small untraceable debris objects (IADC 2007; UN COPUOS. 2010). The UN also advocates that if available orbital data indicates a potential collision during operation, an avoidance manoeuvre should be considered (UN COPUOS. 2010). This requires equipping satellites with the capability to perform collision avoidance manoeuvres as a design consideration.

9.2.3.2 Space Debris Remediation

A growing field of technology research aspires to develop new capabilities to remove space debris objects from orbit (Pelton 2015). The concept of active debris removal (ADR) is similar to post-mission disposal, but requires an external vehicle to supply the mechanism by which the disposal is performed. Simulated studies have demonstrated that ADR of 2 to 5 large debris objects per year has the potential to curb the growth of debris objects by reducing the occurrence of collisions in LEO (Liou 2011; Liou and Johnson 2009). Despite promising research outcomes, ADR technologies are not yet available and there is no clear perspective on how to finance ADR operations at an international level (Bonnal et al. 2013). Nevertheless, ESA continues to pursue what will most likely be "the world's first active debris removal mission" which proposes to launch in 2023 using either a net or robotic arm to capture and deorbit an ESA-owned derelict satellite in LEO.

9.3 Current and Emerging Space Sector Challenges

9.3.1 New Technologies

Space operations have recently shifted from being a largely government-driven to an increasingly commercially-driven activity. Historically, space operations in LEO had been commercially infeasible due to infrequent launches and high capital requirements for satellite design. In the past decade however, commercial interest has grown exponentially as a result of a variety of technological innovations that have reduced the cost of space missions (Besha and MacDonald 2016). The miniaturisation of satellites is one example, a satellite launch mass of 20,000 kg was common in the 1970s while satellites less than 4 kg (cubesats) are frequently being launched today (Datta 2017). Smaller vehicles also enable the disaggregation of satellite services which refers to the use of multiple smaller satellites in place of one larger one (Whitesides 2017). With its fleet of more than 100 Earth Observation Satellites (Planet Labs Inc. 2019), Planet exemplifies commercial actors benefiting from satellite miniaturisation technology. "...The small, standardized size of the cubesat allows us to build a larger, risk tolerant disaggregated sensor network to collect imagery of the entire Earth every day" (Planet representative, as reported by Besha and MacDonald 2016).

Miniaturisation has also catalysed launch frequency, with the Indian Space Research Organisation (ISRO) launching a record-breaking 104 satellites from a single launch in February 2017 (ISRO 2017). Of the satellites launched, Cartosat-2 was the primary satellite along with 2 technology demonstration satellites owned and operated by the ISRO. The remaining 101 satellites launched by the ISRO were owned by international customers (ISRO 2017) including Planet (Datta 2017), taking advantage of the new 'ride share' business model.

SpaceX is another example of a commercial space actor bringing innovation to the space sector to increase participation. Among many other achievements, they were the first to conduct a propulsive vertical landing of a rocket in December 2015 (Grush 2015) and re-launch a refurbished Falcon 9 rocket in March 2017 (Grush 2017). Other than being amongst the most prolific commercial launch providers,

SpaceX is also making significant contributions in the development of satellite technologies for global broadband. At the time of writing, their most recent Starlink launch on 13 June 2020 added an additional 58 Starlink broadband satellites to their constellation. The successful launch extends SpaceX's record as owner of the largest commercial satellite network with a total of 538 Starlink satellites on orbit (Clark 2020), and plans to launch an additional 1,000 Starlink satellites later this year to deliver global broadband services.

Miniaturisation of satellites, reduced cost and availability of components, ride-sharing opportunities, and re-usable launch vehicles, are all new technologies contributing to the rapid growth we are seeing in the "NewSpace" industry (Whitesides 2017). Emerging technologies such as in-orbit satellite servicing and active debris removal, will continue to create growth in the commercial space sector with new economic and business models, lowering technology barriers and costs to encourage new participants and provide new space-based services (Hall 2016). It will be necessary to develop rules that maintain a safe and predictable space environment while at the same time encouraging commercial innovation and growth. This will require states to develop innovative national space laws to support emerging space technologies ranging from global broadband to satellite servicing and asteroid mining.

9.3.2 Governance

Many scholars hold the view that despite the UN COPUOS being at the forefront in development of guidelines to govern the use of space, as a consultative body its policies are non-binding and hence the UN has limited ability to intervene with operations when policies are breached (Durrieu and Nelson 2013; Pace 2015; Williamson 2012). In order for UN policies to be effective, Williamson (2012) states that policies must be incorporated into the rules and regulations that govern space activity at a national level. The Australian licensing framework for space activities requires a space debris mitigation plan that references an internationally recognised guideline, but an analysis by Griffin et al. (2019) notes this is problematic as the UN COPUOS guidelines are not technical and measurable standards that can be enforced. Griffin et al. (2019) points out there are no consequences for non-compliance with the space debris mitigation strategy, despite non-compliance in other areas of the Australian regulations attracting civil penalties. While they recognise the importance of non-binding guidelines in overall space debris mitigation efforts, the authors propose that introducing clear enforcement mechanisms will be necessary to ensure Australian space actors comply with debris mitigation plans. Newman et al. (2020) further draws attention to the fact that launching states have an international obligation to ensure the activities they authorise are compliant with the Outer Space Treaty, but notes that history shows that despite this obligation, launching states have not been held accountable and there are no clear mechanisms by which this can be achieved.

Pace (2015) also holds the view that non-binding guidelines should be incorporated into national regulation, and elaborates that governments need to ensure that domestic regulatory policies keep up with the ever-expanding space capabilities within the private sector. Ideally, regulatory policies should anticipate technological advancement to keep ahead. However, as Pace (2015) notes, there are already regulatory gaps with US licensing processes currently lacking detail for new space technologies. Research toward new technological solutions for rendezvous and proximity operations (RPO), for example, are well underway with applications in space debris removal, on-orbit satellite servicing and asteroid mining. An initiative led by The Consortium for Execution of Rendezvous and Servicing Operations (CONFERS) aims to develop non-binding guidelines for these activities (Newman et al. 2020; Pelton 2019), that may inform a legal framework in the future.

As RPO missions undoubtedly have applications in the space security domain, operations will require enhanced SSA capability to support the technical requirements of intended proximity operations with greater tracking accuracies, and also to distinguish deliberate RPO activities from unintended events such as accidental collisions (Pelton 2019). International SSA data sharing, and the policy mechanisms that enable this, will be critical to accommodate planned activities and aleviate misunderstandings relating to the intentions of national RPO activities. Development of an international legal framework to address issues around the use of these 'dual use' RPO technologies is a challenge that is yet to be resolved, but will be necessary to enable growth and innovation in RPO-related industries.

Also concerning is the lack of information to inform the sustainable operations of global broadband constellations and micro-satellite technology, which was acknowledged by the Inter-Agency Debris Coordination Committee (IADC) in a statement issued in February 2016. This highlights the need for further study of current legislation to identify gaps in regulation. Williamson (2012) explains that in general, approaches to sustainable space operations can be separated across three categories: technological approaches, policy approaches, and legal approaches. He argues that a balance between all three approaches, as well as consideration of economic factors, is necessary toward reaching an international solution for sustainable space operations.

When considering the role of space policy Bradley and Wein (2009) emphasise that the primary focus should be on increasing compliance. Compliance to the "25-year rule" of post-mission disposal has being recognised as "the first defence against the orbital debris population increase" (Liou and States 2015) by numerous studies investigating the growth of the LEO orbital debris environment over a period of 200 years (Klinkrad 2015; Liou and Johnson 2008; Liou 2011; Pelton 2015). The authors' formed this stance in response to their study which determined that the lifetime risk from space debris could be sustained at tolerable levels if the 25-year spacecraft deorbiting guidelines (put forward by the UN COPUOS) were achieved within the next few decades.

Bradley and Wein also provide several detailed, calculated examples, to transform the damages resulting from debris-generating activities due to non-compliance into appropriate fees or taxes – a concept taken from the "polluter pays" strategy implemented by many countries in response to global climate change. Their work also demonstrates the value of drawing from concepts of environmental sustainability on Earth and finding applications for sustainability in space, which is a theme Williamson (2012) proposes may be beneficial toward achieving a condition of sustainable space operations.

9.4 Graph Databases

Existing space object catalogues reflect the primary motivation of SSA by containing data specific to tracking objects in orbit. Already we have a large network of radars and optical telescopes operated by government, military and commercial entities collecting large volumes of data represented in various SSA catalogues. However, these data are not necessarily shared openly with satellite operators, or they exist in fragmented public databases, or expensive propriety catalogues and SSA systems. When reflecting on the current and emerging challenges relating to STM, it becomes clear that achieving internationally coordinated 'safety, stability and sustainability' in outer space cannot rely on SSA data alone. While STM should utilise information from SSA sensor networks, it is also necessary to build upon existing capabilities to include information related to operational practices, policies and regulation.

Graph databases provide the best framework to represent and query complex and dynamic relationships in diverse datasets (Robinson et al. 2015). Today they are widely implemented in commercial and business applications, and have had significant impact by diversifying the types of data captured and queried to generate new insight supporting decision making and strategy. For the same reasons, it is expected that graph database-driven data discovery will have a significant impact when applied to STM.

The general-purpose and expressive structure of graph databases makes them an appealing choice when faced with a complex real-world problem. In contrast to relational databases consisting of tabular data with a fixed structure (the most common format for existing space object catalogues), graphs are made up of nodes connected by relationships. In a labelled property graph, the most widely used graph database model (Robinson et al. 2015), nodes can have multiple labels, relationships are named and directed connecting two nodes only, and both nodes and relationships can have multiple properties (Fig. 9.2).

Figure 9.3 provides a high level graphical representation of how the graph database concept can be applied to support information management and analytics to generate insight for STM problems. The 'Data' represented in Fig. 9.3 includes measurement data from sensors, launch data, satellite data, and any number of other relevant sources. Data is collected, processed, organised and given context to become meaningful 'Information'. These sources of information could include existing space object catalogues, space weather services, launch schedules, space object registries, etc. By understanding how these disparate sources of information

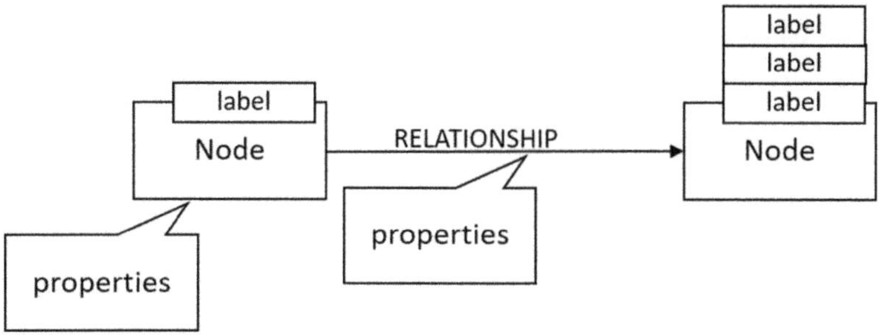

Fig. 9.2 Labelled property graph example showing nodes, relationships, labels and properties

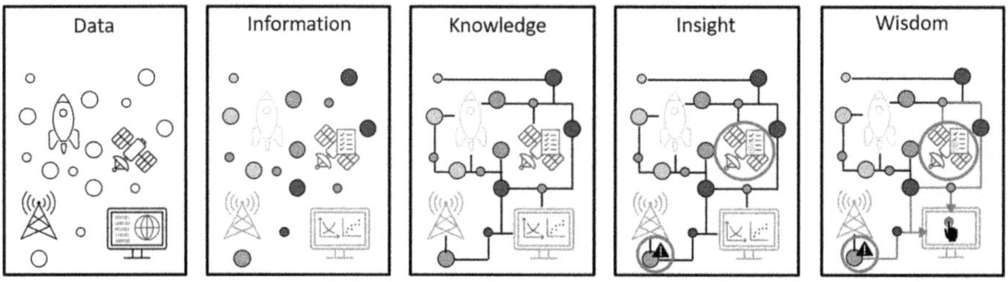

Fig. 9.3 Graph database concept applied to Space Traffic Management. (Illustration adapted from the original by The Gapingvoid. 2019)

are related, and modelling their relationships using a graph database, a 'Knowledge' graph is created. The graph database can then be queried and analysed to generate 'Insight', with pattern recognition and analytics over time creating 'Wisdom' in the space domain.

For example, consider a future agency responsible for monitoring international compliance to space debris mitigation guidelines. Applying the concept in Fig. 9.3, the graph database contains the relevant national and international regulatory frameworks specific to satellites on the basis of their ownership. This information is connected with orbital data from space object catalogues, and querying the graph database can verify an active post-mission disposal manoeuvre in accordance with specific space debris mitigation policies. Over time, identifying patterns of behaviour through the relationships of connected information generate 'Wisdom' to enhance the predictability of space activities and inform international best practice and standards.

Initial research has investigated the implementation of graph databases as a "Space Object Situation Information Service" with a focus on SSA (Lan et al. 2020), and applied graph databases to quantitatively analyse international adherence to the Convention on Registration of Objects Launched into Outer Space (Le May et al. 2020). Future progress toward the development of graph databases to connect the vast amounts of information relating to STM is likely to benefit a broad range of space activities in an increasingly challenging operating environment. Such a tool will be valuable to inform current and future requirements for laws, policy and regulation, while monitoring international obligations related to the stable, secure and sustainable use of space as an international resource.

9.5 Conclusion

This chapter provided an overview of space objects, existing approaches to STM, current and emerging challenges for the space sector, and the potential of graph databases as an analytic tool and knowledge system to support progress toward stable, secure and sustainable space operations in the future. A number of challenges for the space sector were identified, including the rapid growth of commercial space activities, existing instability due to the space debris population, and new technologies such as RPO that provide promising applications to address these challenges – but will require new approaches to space governance and regulation. Overall, this chapter highlighted issues pertinent to STM, and considerations that need to be made for STM as an international challenge that intersects both the technical and governance aspects of operating in a complex and dynamic environment. Graph databases are a technology that can be applied to expand the diversity of information represented in space catalogues, and better unify efforts across technical and policy fields to address STM problems. While initial research has demonstrated some implementations of graph databases in these areas, future work should expand on graph database applications to focus on critical space sector needs including improved SSA capability, the international coordination of space activities, and monitoring compliance to existing space law, policy and regulation.

References

AESB. (1995). *Orbital Debris: A Technical Assessment*. https://doi.org/10.17226/4765.

Anselmo, L., & Pardini, C. (2016). Ranking upper stages in low Earth orbit for active removal. *Acta Astronautica*, *122*, 19–27. https://doi.org/10.1016/j.actaastro.2016.01.019.

Antoni, N., Giannopapa, C., & Schrogl, K. U. (2020). Legal and Policy Perspectives on Civil–Military Cooperation for the Establishment of Space Traffic Management. *Space Policy*, (40), 101373. https://doi.org/10.1016/j.spacepol.2020.101373.

Besha, P., & MacDonald, A. (2016). *Economic Development of Low Earth Orbit*, Washington, DC. https://www.nasa.gov/sites/default/files/atoms/files/economic-development-of-low-earth-orbit_tagged_v2.pdf%20. https://wwwnasa.gov/connect/ebooks/economic_development_of_low_earth_orbit_detail.html.

Bonnal, C., Ruault, J. M., & Desjean, M. C. (2013). Active debris removal: Recent progress and current trends. *Acta Astronautica*, *85*, 51–60. https://doi.org/10.1016/j.actaastro.2012.11.009.

Bradley A. M., & Wein, L. M. (2009). Space debris: Assessing risk and responsibility *Advances in Space Research 43*(9), 1372–1390. https://doi.org/10.1016/j.asr.2009.02.006.

Clark, S. (2020). Starlink satellite deployments continue with successful Falcon 9 launch. https://spaceflightnow.com/2020/06/13/starlink.satellite.deployments-continue-with-successful-falcon-9-launch/.

Datta, A. (2017). The New Space Revolution: The emerging commercial space industry and new technologies. https://www.geospatialworldnet/article/emerging-commercial-space-industry-newtechnologies/.

Diakakis, M., Deligiannakis, G., Katsetsiadou, K., & Lekkas, E. (2015). Hurricane Sandy mortality in the Caribbean and continental North America. *Disaster Prevention and Management*, *24*(1), 132–148. https://doi.org/10.1108/DPM-0520140082.

Durrieu, S., & Nelson, R. F. (2013). Earth observation from space – The issue of environmental sustainability. *Space Policy 29*(4), 238–250. https://doi.org/10.1016/j.spacepol.2013.07.003.

ESA. (2012). ESA declares end of mission for Envisat. https://www.esa.int/Applications/Observing_the_Earth/Envisat/ESA_declares_end_of_mission_for_Envisat.

ESA. (2019). ESA's fragmentation event database statistics. https://fragmentation.esoc.esa.int/home/statistics.

ESA. (2020). Space Environment Statistics. https://sdup.esocesaint/discosweb/statistics/.

EU. (2004). European Code of Conduct for Space Debris Mitigation. European Union. http://www.unoosa.org/documents/pdf/spacelaw/sd/2004.B5-10.pdf.

Griffin, A., Mackenzie, P., Sherborne, A., Sit, H., & Jackson, L. (2019). The Law of Rules: The use of rulese for the advancement of debris mitigation strategies in space activities (D. Blake, Ed.). In D. Blake (Ed.), *Project asteria 2019: Space debris, space traffic management & space sustainability*. Air Power Development Centre.

Grush, L. (2015). SpaceX successfully landed its Falcon 9 rocket after launching it to space. https://www.theverge.com/2015/12/21/10640306/spacex-elon-musk-rocket-landing-success.

Grush, L. (2017). Watch SpaceX relaunch its Falcon 9 rocket in world first The Verge. https://www.theverge.com/2017/3/31/15135304/spacex-launch-video-used-falcon-9-rocket-watch.

Hall, L. (2016). *Opportunities to Advance 'Tipping Point' & Emerging Space Technologies*, (tech. rep.). https://www.nasagov/directorates/spacetech/feature/opportunities-to-foster-commercial-space-technologies-2015.

Harvey J. T. (2002). Estimating census district populations from satellite imagery: Some approaches and limitations. *International Journal of Remote Sensing*, *23*(10), 2071–2095. https://doi.org/10.1080/01431160110075901.

Hughes, B. P., Anund, A., & Falkmer T. (2016). A comprehensive conceptual framework for road safety strategies. *Accident Analysis and Prevention 90* 13–28. https://doi.org/10.1016/j.aap.2016.01.017.

IADC. (2007). IADC Space Debris Mitigation Guidelines, (Revision 1), 1–10. http://www.unoosa.org/documents/pdf/spacelaw/sd/IADC-2002-1-IADC-Space_Debris-Guidelines-Revision1.pdf.

ISRO. (2017). PSLV-C37 Successfully Launches 104 Satellites in a Single Flight – ISRO. https://www.isro.gov.in/pslv-c37-successfully-launches-104-satellites-single-flight.

ITU. (2019). *The State of Broadband : Broadband as a Foundation for Sustainable Development* https://doi.org/10.1017/CBO9781107415324.004.

Jasani, B., Pesaresi, M., Schneiderbauer S., & Zeug, G. (2009). *Remote sensing from space: Supporting international peace and security* (B. Jasani, M. Pesaresi, S. Schneiderbauer, & G. Zeug, Eds.; Vol. 53). Dordrecht: Springer Netherlands. https://doi.org/10.1007/978-1-4020-8484-3.

Johnson, N. L., Krisko, P. H., Liou, J. C., & Anz-Meador P. D. (2001). NASA's new breakup model of EVOLVE 4.0. *Advances in Space Research*, *28*(9), 1377–1384. https://doi.org/10.1016/S0273-1177(01)004239.

Kaiser S. A. (2015). Legal and policy aspects of space situational awareness. *Space Policy*, *31*, 5–12. https://doi.org/10.1016/j.spacepol.2014.11.002.

Klinkrad, H. (2006). Space debris, 5–57. https://doi.org/10.1002/9780470686652.eae325.

Klinkrad, H. (2015). *Space Debris Models and Risk Analysis* (J. Mason, Ed.). Chichester: Praxis Publishing Ltd. https://doi.org/10.1017/CBO9781107415324.004.

Lan, C., Lu, W., Xu, Q., Zhou, Y., Shi, Q., & Lyu, L. (2020). Construction of space object situation information service based on knowledge graph. *IEEE Access*, *8*, 22625–22640. https://doi.org/10.1109/ACCESS.2020.2969704.

Le May S., Carter B. A., Gehly S., Flegel, S., & Jah, M. (2020). Representing and querying space object registration data using graph databases. *Acta Astronautica*, *173*(March), 392–403. https://doi.org/10.1016/j.actaastro.2020.04.056.

Liou, J. C. (2011). An active debris removal parametric study for LEO environment remediation. *Advances in Space Research*, *47*(11), 1865–1876. https://doi.org/10.1016/j.asr.2011.02.003.

Liou, J. C., & Johnson, N L. (2008). Instability of the present LEO satellite populations. *Advances in Space Research*, *41*(7), 1046–1053. https://doi.org/10.1016/j.asr.2007.04.081.

Liou, J. C., & Johnson, N. L. (2009). A sensitivity study of the effectiveness of active debris removal in LEO. *Acta Astronautica*, *64*(2–3), 236–243. https://doi.org/10.1016/j.actaastro.2008.07.009.

Liou, J., & States, U (2015). The InterAgency Space Debris Coordination Committee (IADC) An overview of the IADC annual activities Overview of IADC • The IADC is an international forum of national and, (February), 1–13.

Matney M., & Settecerri, T. (1997). Characterization of the breakup of the pegasus rocket body 1994-029B, In *Second european conference on space debris*, Darmstadt, European Space Agency. https://doi.org/10.16309/j.cnki.issn.10071776.2003.03.004.

Mcnally T., Bonavita, M., & Thépaut, J. N. (2014). The role of satellite data in the forecasting of hurricane sandy. *Monthly Weather Review*, *142*(2), 634–646. https://doi.org/10.1175/MWRD1300170.1.

NASA Orbital Debris Program Office. (2020). *Orbital Debris Quarterly News*(tech. rep. No. 2). Houston. https://doi.org/10.1016/00945765(93)90151-L.

Newman, C., Dinsley R., & Ralston, W. (2020). Introducing the law games: Predicting legal liability and fault in satellite operations. *Advances in Space Research*, https://doi.org/10.1016/j.asr.2020.04.020.

Newman, L. K., Frigm, R., & Mckinlev D. (2009). It's not a big sky after all: Justification for a close approahc prediction and risk assessment process (A. V. Rao, T. A. Lovell, F. K. Chan, & L. A. Cangahuala, Eds.). In A. V. Rao, T. A. Lovell, F. K. Chan, & L. A. Cangahuala (Eds.), *Advances in the astronautical sciences*. Pittsburgh: American Astronautical Society.

Newman, L. K., Frigm, R., & McKinley D. (2010). It's not a big sky after all: Justification for a close approach prediction and risk assessment process. *Advances in the Astronautical Sciences, 135* 1113–1132.

Nicolls, M., Vittaldev V., Ceperley D., Creus-Costa, J., Foster C., Griffith, N., Lu, E., Mason, J., Park, I.-K., Rosner C., & Stepan, L. (2017). Conjunction Assessment for Commercial Satellite Constellations Using Commercial Radar Data Sources. *Advanced Maui Optical and Space Surveillance Technologies Conference (AMOS)* www.amostech.com.

Pace, S. (2015). Security in space. *Space Policy*, *33*(December 2013), 51–55. https://doi.org/10.1016/j.spacepol.2015.02.004.

Pardini, C., & Anselmo, L. (2009). Assessment of the consequences of the Fengyun-1C breakup in low Earth orbit. *Advances in Space Research*, *44*(5), 545–557. https://doi.org/10.1016/j.asr.2009.04.014.

Pardini, C., & Anselmo, L. (2011). Physical properties and long-term evolution of the debris clouds produced by two catastrophic collisions in Earth orbit. *Advances in Space Research*, *48*(3), 557–569. https://doi.org/10.1016/j.asr.2011.04.006.

Pelton, J. N. (2013). *Space Debris and Other Threats from Outer Space* https://doi.org/10.1007/9781461467144.

Pelton, J. N. (2015). *New Solutions for the Space Debris Problem*. Cham: Springer International Publishing. https://doi.org/10.1007/978-3-319-17151-7.

Pelton, J. N. (2019). A path forward to better space security: Finding new solutions to space debris, space situational awareness and space traffic management. *Journal of Space Safety Engineering*, *6*(2), 92–100. https://doi.org/10.1016/j.jsse.2019.04.005.

Pham, T. T. H., Apparicio, P., Gomez, C., Weber C., & Mathon, D. (2014). Towards a rapid automatic detection of building damage using remote sensing for disaster management: The 2010 Haiti earthquake. *Disaster Prevention and Management: An International Journal*, *23*(1), 53–66. https://doi.org/10.1108/DPM-12-2012-0148.

Planet Labs Inc. (2019). Our Constellation. https://storage.googleapis.com/planet-ditl/day-in-the-life/index.html.

Reynolds, R. C., Eichler P., Bade, A., Krisko, P., & Johnson, N. L. (1999). Sensitivity analysis of the orbital debris environment using the evolve 4.0 model. *Advances in Space Research*, *23*(1), 175–185. https://doi.org/10.1016/S02731177(99)00002-2.

Robinson, I., Eifrem, E., & Webber, J. (2015). *Graph Databases: New Opportunities for Connected Data* (Beaugureau (ed.); 2nd ed.). O'Reilly Media.

Rosanelli, R. (2014). Goodbye Envisat, or legal issues associated with the threat of a massive space debris. http://www.spacesafetymagazine.com/space-debris/kessler-syndrome/goodbye-envisat-legal-issues-threat-massive-space-debris/.

Scientific and Technical Subcommittee. (1999). *Technical Report on Space Debris*.

Skinner M. A., Jah, M. K., McKnight, D., Howard, D., Murakami, D., & Schrogl, K. U (2019). Results of the international association for the advancement of space safety space traffic management working group. *Journal of Space Safety Engineering*, *6*(2), 88–91. https://doi.org/10.1016/j.jsse.2019.05.002.

Stilwell, R. E., Howard, D., & Kaltenhauser S. (2020). Overcoming sovereignty for space traffic management. *Journal of Space Safety Engineering*, *7*(2), 158–161. https://doi.org/10.1016/j.jsse.2020.04.002.

The Gapingvoid. (2019). Want to know how to turn change into a movement? https://www.gapingvoidcom/blog/2019/3/5/want-to-know-how-to-turn-change-in-to-a-movement/.

UN COPUOS. (2010). *Space Debris Mitigation Guidelines of the Committee on the Peaceful Uses of Outer Space* (tech. rep.). https://doi.org/A/RES/62/217.

United Nations General Assembly. (2015). *Transforming our world: The 2030 agenda for sustainable development* (tech. rep. No. 1), New York. https://doi.org/10.1007/s13398-014-0173-7.2.

Walker R., Martin, C. E., Stokes, P H., Wilkinson, J. E., & Klinkrad, H. (2001). Analysis of the effectiveness of space debris mitigation measures using the DELTA model. *Advances in Space Research*, *28*(9), 1437–1445. https://doi.org/10.1016/S02731177(01)004458.

Walker R., Stokes, P H., & Wilkinson, J. E. (2000). 0094-5765/00 $ – see front matter, *47*, 707–717.

Whitesides, G. (2017). Where will space technology take us by 2030, and what does this mean for life on Earth? | World Economic Forum. https://www.weforum.org/agenda/2017/02/space/.

Wiedemann, C., Oswald, M., Stabroth, S., Klinkrad, H., & Vörsmann, P (2005). Modeling of RORSAT NaK droplets for the MASTER 2005 upgrade. *Acta Astronautica*, *57*(2–8), 478–489. https://doi.org/10.1016/jactaastro2005.03.014.

Williamson, R. A. (2012). Assuring the sustainability of space activities. *Space Policy*, *28*(3), 154–160. https://doi.org/10.1016/j.spacepol.2012.06.010.

World Health Organization (WHO). (2017). *A Road Safety Technical Package* (tech. rep.). http://ris.paho.org/xmlui/bitstream/handle/123456789/34980/9789275320013-por.pdf?sequence=1&isAllowed=y.

Chapter 10
Space Debris, Space Situational Awareness and Cultural Heritage Management in Earth Orbit

Alice Gorman

Abstract Since the launch of Sputnik 1 in 1957, Earth orbit has become filled with human-made debris. It is widely recognised that some form of active debris removal is needed to secure ongoing access to space. The dominant paradigm for approaching space debris is Space Situational Awareness (SSA), which foregrounds the instrumental values of Earth orbit. However, numerous defunct satellites in Earth orbit have cultural significance for populations of Earth for their historic, aesthetic, scientific and social values, which are not easily accommodated within the SSA model. I argue that effective heritage management is better pursued using an environmental management framework which incorporates the intrinsic values of the space environment as a new entity created by the synergistic effects of natural and cultural activities. For heritage satellites, protection in situ is an appropriate management strategy when the collision risk is low. This paper elaborates on mechanisms and concepts, such as heritage lists, themes and offsets, necessary for ensuring that some culturally significant spacecraft survive into the future.

10.1 Introduction

Space debris is recognised as a growing problem for space industry and science due to the cumulative effects of collisions in orbit (Klinkrad 2015). Currently, it is estimated that after approximately 7000 launches since 1957, there are over 34,000 objects greater than 10 cm in size in Earth orbit (NASA 2020). Of these, approximately 6% are operational craft, with the remainder being defunct spacecraft and fragmentation debris (European Space Agency 2020). The region between 200 and 1200 km is densest in debris, with satellites, upper rocket stages, space stations, and millions of fragments. In weight, space debris is estimated to be 8000 tons, or the equivalent of ten million cane toads.

A. Gorman (✉)
College of the Arts, Humanities and Social Sciences, Flinders University, Adelaide, SA, Australia
e-mail: Alice.Gorman@flinders.edu.au

M. de Zwart, S. Henderson (eds.), *Commercial and Military Uses of Outer Space*, Issues in Space, https://doi.org/10.1007/978-981-15-8924-9_10

Cane toads were introduced to Australia in the 1930s to control sugar cane pests. They are an apt analogy as, like orbital debris, they are now noxious and out of control (Gorman 2019a). All the same, there are some frog princesses in this amphibian morass that merit further attention for their cultural heritage value.

On Earth, cultural heritage management frequently takes place within an environmental management framework. As I have argued elsewhere (Gorman 2005a), there is no fully developed concept of environment for space. The absence of life, coupled with the terrestrial tendency to privilege ecosystems, has left the space environment as a deficit rather than a presence. In general, space researchers have failed to draw on the extensive scholarship and practice of terrestrial environmental management. Instead, management of orbital debris is considered part of the military-derived concept of Space Situational Awareness (SSA). In this chapter I investigate how heritage operates within the paradigms of 'environment' and 'situation', before identifying specific steps needed to develop off-Earth heritage management. These include the creation of heritage lists, the development of heritage themes, and mitigation strategies.

10.2 Background to Space Heritage

While the public generally associates heritage with the deeper past, particularly with monumental architecture and the built environment, the simple definition of heritage is 'things from the past and present, worth preserving for present and future generations' (Aplin 2002). The present is where decisions get made about what is sent forward into the future (although the validity of this future-leaning project has been critiqued, for example by Högberg et al. 2017). Increasingly, there is a recognition that the material culture of the contemporary world, including the Cold War, aviation, computing, and space exploration, is worthy of heritage protection (Harrison and Schofield 2010).

The heritage of space exploration is embedded in landscapes on Earth, where the terrestrial infrastructure related to the development, launch, tracking and return of spacecraft is located. Off-Earth, we find satellites, spacecraft, and space junk in Earth orbit; landing sites on planets, moons, asteroids and comets; and spacecraft and probes ranging from close to the Sun to interstellar space (Capelotti 2010). These places and objects are the tangible heritage of the human movement into space, but there are also the intangible associations, practices and beliefs related to contemporary space activities. Together they form what we might call a 'space culture' (Capelotti 2010; Gorman 2016a).

The old satellites and millions of fragments of spacecraft in Earth orbit are part of the material record of this space culture, and have their own distinct heritage issues. It's clear that active debris removal is required to ensure continued access to and use of this space; but this must be balanced against the cultural significance of space junk. In broad terms, space junk demonstrates the development of technologies enabling humans to engage directly with space. These technologies have been transformative of human experiences of life on Earth, as everyday activities have come to depend on space-delivered services such as navigation, timing, telecommunications and Earth observation. The documentary record, with all of its biases and lacunae, tells only some of this story. The objects themselves, within their environment, are dynamic rather than frozen on paper; and reveal different facets to the investigator.

The survival of space-related places on Earth is often dependent on national or state heritage legislation, where these recognise the heritage of the recent past (frequently heritage must meet an age requirement such as 100 years). In space, however, the principles of the Outer Space Treaty (OST) sit uneasily with the requirements of terrestrial heritage regimes. Article 2 of the OST forbids territorial claims, making the extension of state-based heritage legislation problematic (O'Leary 2006). Existing international heritage conventions are not a viable alternative at this stage. It is not difficult to make an argument that some pieces of space junk, such as Vanguard 1, the oldest human object in space, have sufficient 'outstanding universal significance' to be considered world heritage. However, the World Heritage Convention does not cover moveable heritage such as ships, which are equivalent to satellites. Inscription on the World Heritage List is dependent on nations initiating and prosecuting nominations: it is inherently nationalist in character (Labadi 2007; Anglin 2008).

Nonetheless, there are many defunct spacecraft in the debris population with a high level of cultural significance for contemporary communities, and their future survival should be considered. At present threats to this heritage derive from factors in the space and gravitational environment, such as aerodynamic drag or collisions with space objects, over which we have little control. What we can control is the implementation of active debris removal strategies.

The aim of cultural heritage management is to retain cultural significance. Despite the seemingly remote location and radically different environment, terrestrial heritage management principles can be applied in space (Gorman 2005a, b, 2015, 2017). Following Australia's justly famous set of heritage guidelines, the Burra Charter, the location in orbit, or the setting, is part of this significance. The apparently easy solution, to remove historic spacecraft to a terrestrial museum when feasible to do so, is not the most appropriate management strategy as it diminishes cultural significance. This seems at odds with the urgent need to 'clean up' space junk by removing objects from orbit. However, the paradox is not intractable. Many significant spacecraft present low collision risks at present. (It must be noted that the population of Low Earth Orbit is currently undergoing rapid change with the launch of Starlink and other massive constellations, eg McDowell 2020).

10.3 Situation vs Environment

At the definitional level, situation and environment describe quite different entities with different value systems. A situation is the way in which something – usually an event, but also objects and people – is related to its surroundings; or, indeed, its environment. A situation can also describe a finely balanced set of circumstances or a critical point in time. A situation is often problematic, requiring solution or resolution.

The delineation of a situation is the foundation of Space Situational Awareness. SSA can be defined as

> the requisite foundational, current, and predictive knowledge and characterization of space objects and the operational environment upon which space operations depend – including physical, virtual, information, and human dimensions – as well as all factors, activities, and events of all entities conducting, or preparing to conduct, space operations (Chairman of the Joint Chiefs of Staff 2018)

The concept of situational awareness derives from military strategy. It is instrumental and teleological in nature: instrumental value is a means to an end, and that end is what is useful to consumers of space services. The application of SSA is aimed at continuing the human exploitation of the affordances of the space environment. Implicit in this formulation are the goals of furthering national security, commercial uses, or services supplied from space. While a human dimension and part of the characterisation of space objects, it is difficult to see how heritage is a dependency for space operations. The weight of public opinion about the destruction of space heritage could conceivably be interpreted as an impediment.

While environment is also 'surroundings', it is not focused on a particular point or goal. It is a complex web of factors which may include physical objects or spaces, climate, biological and chemical entities, mechanical processes, and social circumstances. It is often associated with ecologies. The interdependencies mean that change in one aspect of an environment may affect other aspects. While environments can have instrumental value to humans and other living organisms, they can also have intrinsic value (in contrast to a situation). According to Justus et al. (2009),

> Ascribing intrinsic value to non-human natural entities is intended to underscore reverence for nature by according them a value independent of humans, thereby liberating humans from narrow anthropocentrism about value.

Intrinsic value is inherent in the thing itself rather than the benefits it affords. In the words of Westacott (2019), 'the universe is somehow a better place for that thing existing or occurring'.

In the face of human industrial, agricultural and urban developments, environmental management seeks to minimise or mitigate damage to ecological and social systems. It can be defined as

> actual decisions and action concerning policy and practice regarding how resources and the environment are appraised, protected, allocated, developed, used, rehabilitated, remediated, and restored. (Zurlini et al. 2008)

These actions may be directed towards retaining the instrumental value of particular environments, but equally towards the intrinsic value.

Because human attention is so focused on terrestrial environments which feature life, non-biological environments have not been fully conceptualised. This betrays the shortcomings of our limited terrestrial perspectives when applied to space. But what is this environment, in the absence of biological ecologies in orbit? A concept from heritage may bridge this gap. The nature of the heritage resource in space is not just the aggregation of individual space objects, considered part of the 'situation' in terms of their location and predicted collision risks. It is also the web of relationships between objects, and between objects and their environmental setting. Since the launch of Sputnik 1 in 1957, the near-Earth environment can no longer be considered 'empty' space. The combination of natural and cultural processes at work have refigured it as a unique cultural landscape which has intrinsic value (Gorman 2005a, b, 2009a, 2015). The concept of the site in Earth orbit raises the issue of how the object relates to a place through which it only passes. The cultural landscape provides an alternative avenue for characterising the nature of the relationship between the object and its environment, one which takes account of the contribution of each.

While there are obvious points of overlap between situation and environment, there are also significant differences between the two approaches which lead to different strategies and policies. Dialogues around 'sustainability' in space have centred on SSA rather than environmental management. However, there is a growing acknowledgement, for example in the Artemis Accords (NASA 2020), the Hague 'building blocks' (The Hague International Space Resources Governance Working Group 2019) and the Vancouver Recommendations on Space Mining (Outer Space Institute 2020), that heritage is a factor in the sustainable use of space. Typically, heritage is considered a non-renewable resource (although see Holtorf 2001 for a counter-argument) which contributes to social identity and well-being (eg Power and Smyth 2016).

10.4 The Legal and Regulatory Environment

While heritage principles apply in space, the legal and regulatory environment is distinctly different and presents some interesting challenges. National or state heritage legislation is limited to the protection of objects rather than sites, and has only been tested in the lunar setting (Westwood et al. 2010). In the absence of a heritage regulation regime, international charters and conventions provide some waypoints.

Key to the way heritage management and archaeology operate on Earth are the concepts of 'site', 'place' and 'landscape', all of which incorporate both human-made artefacts and structures, the surrounding environment, and the relationships between them. Soil deposits containing archaeological remains are considered part of the site, as they preserve biological and geological evidence of the past environment in which humans lived at that time and in that place. Following the cessation of human activity, the place continues to change. These post-depositional processes disperse the behavioural connections between artefacts, structures and environment which the archaeologist aims to decipher – a method known as taphonomy.

Under the terms of the Outer Space Treaty, orbital objects are owned by the launching state. The space through which the object moves is unaffected by its presence: ownership is not conferred by use and the presence of the object does not constitute a territorial claim. The OST conceptually isolates human-made objects from their environment, or setting, thus reinforcing the philosophy of SSA.

The 'place' occupied by a piece of space junk is dynamic and is defined by orbital equations rather than static co-ordinates (Gorman 2009b). The local environment is not generally considered to be associated with the artefact or to be part of its setting, perhaps because it appears to have no environmental setting: the myth of the 'vacuum' of space, coupled with the sparseness and invisibility to human senses of its constituents, mitigates against this.

Place in space is, in some senses, a non-place (Augé 1992), characterised by transience and anonymity: somewhere lacking enough 'stickiness' to be a place in human terms. Within this non-place, this invisible setting, the task is to evaluate the level of heritage significance accrued by an artefact over its period of time in orbit. For this, the Burra Charter 2103 is eminently applicable (Gorman 2005a, b, 2015). The Charter was developed in 1976 by the Australian chapter of the International Council on Monuments and Sites (ICOMOS), specifically to address blind spots in existing charters and principles, which were not well adapted to Indigenous places and settler built environments.

The process of significance assessment outlined in the Burra Charter incorporates five categories: historic, aesthetic, scientific, social and spiritual. Although there are other significance schemes, for the most part they are covered by the Burra Charter categories (eg Lipe 1984). Where conflicting meanings exist, the Burra Charter recommends that all are accommodated, rather than privileging a single authorised version. Cultural significance is determined through a process of research and consultation with stakeholders, often but not necessarily accompanied by field recording to ascertain the condition and how much of the significant fabric (material) survives. It is the first step to devising appropriate and practical management strategies. In the space community, it is often assumed that heritage means preservation; however, preservation is just one of a multitude of strategies to mitigate damage to cultural values in the process of development.

The Dublin Principles (2011) on the conservation of industrial heritage sites, structures, areas and landscapes also have relevance for the management of space heritage. The Principles are a joint statement by ICOMOS and the International Committee for the Conservation of the Industrial Heritage (TICCIH). The Principles acknowledge that industrial heritage includes 'complexes and multiple site operations or systems, whose many components are interdependent, with different technologies and historical periods frequently present'. This formulation is particularly apt for space technology which incorporates terrestrial and off-Earth components.

Recommendation II 6 of the Dublin Principles, which states that 'Integrated inventories and lists of structures, sites, areas, landscapes, their setting and associated objects, documents, drawings and archives or intangible heritage should be developed', is a good starting point for the cultural heritage management of space. While there is an enormous amount of documentation relating to space activities, much has also been lost, is poorly curated, or just inaccessible.

National heritage legislation cannot be directly applied to space sites or landscapes but can provide guidance on the processes and administration needed to manage celestial heritage. Useful concepts are heritage themes, heritage lists or registers, requirements for consultation, notification, and levels of heritage – eg local, national, regional, global. Some of these are explored in the next section.

10.5 Assessing Cultural Significance in Orbit

Not all space junk is equal: some will have greater levels of cultural significance than others. The population of debris for which significance can be assessed is dependent on the limits of observation and the ability to ascribe an origin for the debris (incidentally, two factors which also affect the attribution of liability for the purposes of the UN Convention on International Liability for Damage Caused by Space Objects). Generally, objects below 10 cm in size are difficult to identify and track, although there are improvements in data-collecting and modelling happening all the time. Without knowing the source of the debris, a comprehensive significance assessment of a small fragment is difficult to do, although such fragments can be treated as a statistical population. The low-hanging fruit in orbital cultural heritage management is whole or nearly complete spacecraft.

Vanguard 1 (launched in 1958), the oldest satellite still in orbit, Telstar 1 (1962; Fig. 10.1), the first active telecommunications satellite, and Syncom 3 (1964), the first geostationary satellite, are now technically space junk. These historically significant US satellites were all launched within the first decade of space exploration. Table 10.1 outlines, very briefly, their broader cultural significance using the Burra Charter categories (see Gorman 2005b for fuller descriptions). Satellites from the US and USSR/Russia dominate the material culture in Earth orbit: these nations have launched the highest number of spacecraft and contributed the most orbital debris. The story of the Space Race is well known, and also tends to dominate the historical significance of spacecraft. It is possible to argue that these satellites have a global significance for their role in the development of space technology and telecommunications. However, other satellites, narratives, nations and communities are often overlooked in the standard 'Space Race Model' of cultural significance of space artefacts and places (Gorman 2005a), as we see in the example which follows.

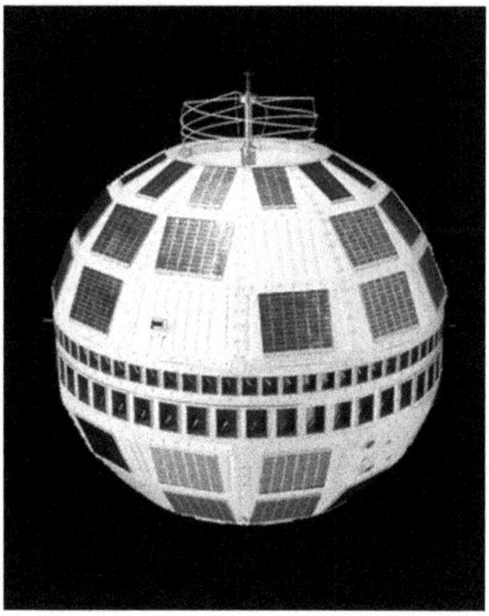

Fig. 10.1 Telstar 1. (Image courtesy of NASA)

Table 10.1 Burra Charter significance categories applied to three satellites

Satellite	Historic	Aesthetic	Scientific	Social
Vanguard 1	Second US satellite; Cold War; international volunteer cooperation in Project Moonwatch; International Geophysical Year	Spherical aluminium sphere; distinctive antennas	Material survival in space for 60 years	Widely known; subject of art projects eg Project Adrift; online fan groups
Telstar 1	First active telecommunications satellite; joint commercial/government	Spherical with distinctive solar panel arrangement	Development of satellite telecommunications technology	Name still used to signify space technology; popular culture such as Adidas soccer ball, Tornados pop single
Syncom 3	First GEO satellite; used in Vietnam War, 1964 Tokyo Olympic Games, associated with growth of domestic television	Cylindrical spinner	Development of satellite telecommunications technology; stabilisation techniques	Unknown

From Gorman (2017)

10.5.1 *Social Significance*

Social significance is a measure of contemporary community esteem for a place or object. Optus B1 (currently in the GEO graveyard orbit) was one of Australia's early telecommunications satellites, launched from China in 1992. This series of satellites was commissioned from the US-based Hughes Space and Communications Company, and Australia had no part in the design. As well as providing regular services such as television, telephone and air traffic control, Optus B1 hosted the ComRadSat channel to service Australia's community radio network. Optus B1 thus has social significance for the thousands of community broadcasters in Australia who used the satellite to downlink content, including feminist, Indigenous and LGBQTI programming (Fredericks 1993). Community radio is an important alternative to commercial and government services, often highlighting marginal voices and views, or regions otherwise poorly serviced. In Australia, community radio has also been influential in establishing an independent Indigenous media, for example, through the Central Australian Aboriginal Media Association (CAAMA).

A consideration is the weighting given to each category of significance. For the assessment of Aboriginal heritage in Australia, for example, social and/or spiritual significance takes precedence over historic or scientific significance in management decisions, in an effort to redress historic injustices in which Aboriginal heritage was erased in a colonial process of genocide. In space, social significance may be less important as orbital places are often less connected to communities due to inaccessibility (although these can coalesce around ground stations such as the Canberra Deep Space Communications Complex at Tidbinbilla). People cannot easily see orbital heritage or interact with it in the same way as terrestrial heritage; it is far less visual than, for example, the Apollo 11 lunar landing site. At this stage, however, no assumptions should be made. Terrestrial heritage management shows that professional and community opinions on social significance are often radically different, and people do form attachments to space hardware which are part of their social significance (Gorman 2016b).

10.5.2 *Scientific Significance*

Scientific significance refers to the research potential of an object or place. Confusingly, this is not about the scientific contributions made by a satellite, which are more properly historic significance. What research questions we would be unable to answer if this artefact or place were destroyed? In order to describe this, it is important to know how *representative* a place or artefact is; ie, how commonly or rarely it occurs. Take the biro, or ballpoint pen, the subject of a joke in The Hitchhiker's Guide to the Galaxy in which all the missing pens are located on a single planet. The destruction or loss of one makes no difference when there are millions of them; but reduce that number to 10, and the same artefact,

Table 10.2 Upper stages in LEO

Name	Number in LEO	Mass in LEO (kg)
Kosmos (SL-8) Second Stage	290	416,150
Zenit-2 (SL-16) Second Stage	22	198,000
Tsiklos (SL-14) Third Stage	110	154,770
CZ-4 Third Stage	33	66,000
Vostok (SL-3) Second Stage	53	58,300
CZ-2C Second Stage	13	52,000
Delta 1 & 2 Second Stage	54	47,375
Dnepr-1 (SL-24) Third Stage	16	37,760
CZ-2D Second Stage	5	20,000
Total	**596 (72.6%)**	**1,050,355 (87.6%)**

From Anselmo and Pardini (2016)

despite its cheapness or disposability, acquires a much greater level of scientific significance. This is exactly the case for computing heritage: once commonplace pieces of hardware are now rarities, as no-one thought they were worth preserving at the time.

The space junk equivalent of the ballpoint pen might be the rocket body, of which there are hundreds abandoned in Low Earth Orbit. They constitute a major risk of fragmentation due to the instability of residual fuel. Although they re-enter the atmosphere frequently, their population is also constantly added to. Table 10.2 shows the most numerous by weight and frequency from 2016 data.

The Kosmos (SL-8) Second Stages are the most numerous, and also contribute the most weight to the cane toad quotient. They are ideal candidates for active removal. However, if scientific significance is taken into account, it becomes obvious that a sample of all classes of rocket body should be retained in situ, if the level of risk were acceptable.

10.6 Heritage Themes

As well as considering the numerical representation of classes of spacecraft, there should be ideally a representation of themes in orbit. The role of heritage themes is to link sites and objects to historical, social and technological processes, in the context of other places or objects that relate to them (eg Heritage Council of NSW 2001). Themes can connect orbital objects through time and across Earth and space, in accordance with the Dublin Principles' emphasis on the connectivity of industrial heritage.

Transportation is a common theme used for terrestrial sites. It includes everything related to roads, rail, ports, ships, aviation, the automotive industry, rockets, space stations and space transportation systems, and cultural practices such as the 'road trip' (Gorman et al. 2018).

As well as transportation, and recognising that a spacecraft may intersect with more than one, other relevant themes include:

- Amateur and citizen science
- Defence
- Cold War
- Telecommunications
- Astronomy and other science
- Earth observation
- Navigation
- International co-operation
- The evolution of space technology
- Propulsion and energy
- Civil and commercial space – developing local, regional and national economies
- Indigenous space and colonialism
- Labour
- Education
- Cultural life – creative endeavours, domestic life, leisure, sport, social institutions

The satellites Telstar 1 and Syncom 3 could be related to the themes of telecommunications, civil and commercial space, evolution of technology and cultural life. They were launched only two years apart, but Telstar 1 represents LEO-based telecommunications, while Syncom 3 marks the move into GEO. This resulted in a change in the ground segments required to support the satellites. Telstar's ground segment consists of six very interesting ground stations: the dish at Goonhilly Downs in Cornwall, still a very active antenna with its own music festival, horn antennas at Andover in Maine, US and Pleumeur Bodou in France, as well as dishes in Raisting, Germany, Fucino, Italy and Nova Scotia, Canada. The mention of Telstar 1 still elicits memories of the instrumental pop hit by the Tornados, the first British act ever to top the US musical charts in 1962 (Philo 2014).

In the master narrative of the Space Race, a frequently forgotten thread is the early tradition of amateur radio satellites. Project Oscar and it successor AMSAT supported amateur groups to build their own satellites and negotiated on their behalf for launch by space powers such as the US. They coordinated volunteer networks across the world to track and report – an inspiring example of international cooperation. These projects are part of a trajectory that includes today's cubesats, which are often student or start-up projects. The fifth Project Oscar amateur radio satellite, Australis Oscar 5, celebrated its 50th anniversary in orbit in 2020. It was designed and built by Melbourne University students in the mid-1960s and launched in 1970 (Gorman 2009c, Mace 2017). This heritage satellite is a touchstone that tells a very Australian technological story about ingenuity and the triumph of the underdog.

Association with a theme assists in the decision-making process about what should be kept in orbit for the future. The connections between these satellites can tell a story with greater geographical and chronological depth than focusing on the individual spacecraft – further demonstration that the cultural landscape of technology in orbit has value in its own right.

10.7 Heritage Lists

The idea of a heritage list of orbital objects was first proposed by Barclay and Brooks (2002). Terrestrial heritage legislation often establishes registers or lists of heritage properties. Registration requires meeting significance criteria appropriate to local, state, national and global legislation or conventions. The sample of heritage places which make it onto a register reflects community values as they change over time. Early lists, such as Japan's Plan for the Preservation of Ancient Artifacts 1871 and the UK's Ancient Monuments Protection Act 1882, tended to focus on grand or stately buildings and the monuments of classical or archaeological civilisations, representing a narrative of colonial power and a linear evolution of cultural superiority culminating in European architecture. This philosophy has been thoroughly critiqued, and in contemporary times, registered places now include more diverse communities and landscapes, including industry and technology.

Typically, registration offers some protection to a heritage place. The administration of heritage legislation may establish requirements to

- obtain permits prior to any alteration or disturbance to a heritage place;
- prepare a conservation management plan, which outlines actions to conserve the fabric;
- prepare a cultural heritage management plan, which outlines processes to protect cultural significance during development activities in the locality;
- consult with stakeholders (Gorman 2017).

None of these measures may be sufficient to protect a heritage place against commercial or industrial developments, when other competing interests are supported by governments. Registers have numerous issues: they are often incomplete or represent specific ideologies, information is often lacking, and they can rapidly become out-of-date. The establishment of a list is only the first step. It also requires the allocation of resources and dedicated administration.

The problem for space heritage is that the inclusion of an off-Earth site on a national, state, or local heritage register potentially contravenes the OST. However, a space object remains the property of the launching state under Article VIII. In 2010, Westwood et al. (2010) were successful in nominating the objects abandoned at Tranquility Base on the Moon to the California State Register of Historical Resources as well as the New Mexico State Register of Cultural Properties. Only the artefacts, and not the site or place as understood in heritage terms, were listed. While important symbolic recognition for the historical significance of the site, this registration does not extend to the site itself. As I have argued (Gorman 2019a), the site is more complex even than the simple equation of artefacts + environment: it also includes shadows and the distinctive microenvironment created by the levitation of lunar dust as the day passes over it.

Another precedent is provided by Australia's List of Overseas Places of Historic Significance to Australia, proposed by Prime Minister John Howard in the early

2000s as part of the Environmental Protection and Biodiversity Conservation Act 1999. The List was adopted in 2007. Its purpose was to

> symbolically recognise sites of outstanding historic significance to Australia located outside of the Australian jurisdiction ... in a way that is respectful of the rights and sovereignty of other nations. (Commonwealth of Australia 2017)

Only three properties have ever been inscribed on this list. They are Anzac Cove in Gallipoli, Turkey, a foundational place in myths of Australian nationhood, the World War II Kokoda Track in New Guinea, and Howard Florey's Laboratory in Oxford, associated with the development of penicillin. The list has no legal standing to protect the places as Australian jurisdiction obviously does not extend to other nations; but it lends 'moral weight'. The creation of this list was a political act to bolster a nationalist narrative, but this aside, the principle seems applicable to space.

Heritage lists maintained by Non-Governmental Organisations (NGOS) are another matter. They are not backed by legislation and have the capacity to inscribe places across national boundaries. The disadvantage is that they fail to provide any further protection than 'moral weight'. Barclay and Brooks (2002) proposed establishing a Commission under the auspices of the International Union for the History and Philosophy of Science and Technology, which could operate in a similar fashion to UNESCO. This suggestion was not pursued at the time, perhaps because, nearly 20 years ago, the problem of space junk did not have the urgency it has now, and the concept of space heritage was so new (the first conference session on this topic was at the World Heritage Congress in 2003).

The American Institute of Aeronautics and Astronautics' (AIAA) list of Historic Aerospace Sites is perhaps the only one that includes off-world heritage places as well as terrestrial. Tranquility Base on the Moon is listed. Established in 1999, the aim of the programme was 'to promote the preservation and dissemination of knowledge about significant accomplishments of the aerospace profession" (AIAA n.d.). As the AIAA has members in many countries, this list is important because it represents the values of the international aerospace community.

Many look to the World Heritage List, administered by UNESCO, as a solution to space heritage; but its limitations at present make its application to orbit problematic. In any case, most satellites would not meet the criterion of outstanding universal value. Other options should be explored, particularly as threats to space heritage both in orbit and on the Moon are accelerating. UNESCO does have a role to play outside the specific context of world heritage. A UNESCO space heritage list could be supported by the International Council on Monuments and Sites (ICOMOS), which advises UNESCO on heritage matters and sets international principles and norms. An ICOMOS Scientific Committee to consider space heritage was initiated by the Australian chapter in 2007 following the Extreme Heritage conference in Cairns but did not continue. There is goodwill towards reviving this committee, which would be in a position to develop a Charter on space heritage, including orbital debris, and outline the principles needed to develop the criteria for an indicative list.

The UN Register of Objects Launched into Outer Space, established in 1962, and maintained by UNOOSA, was proposed by Fewer (2002) as the basis for an orbital heritage list. The Register currently contains over 9500 orbital objects, some under the terms of the Convention on Registration of Objects Launched into Outer Space (1976), and others unregistered. As the infrastructure of the register is already in place, and the basic orbital status of the objects recorded, it would seem a straightforward process to add layers of heritage information such as significance assessment and condition reports, along the lines of terrestrial heritage registers.

Given the dynamic status of orbital heritage, there is a solution which offers even more in terms of active heritage management. Databases which are used for collision and re-entry analysis specify the risks presented by and to a heritage object in more-or-less real time. Conjunction analyses act as a notification mechanism if a collision is predicted. Several of these databases exist. The European Space Agency maintains DISCOS (Database and Information System Characterising Objects in Space), which is

> a single-source reference for launch information, object registration details, launch vehicle descriptions, as well as spacecraft information (e.g. size, mass, shape, mission objectives, owner) for all trackable, unclassified objects (ESA n.d.).

Other international NGOs who could take on a coordinating role are the Committee for Space Research (COSPAR) and the Inter-Agency Debris Coordination Committee (IADC). The graph database developed by LeMay (this volume) offers innovative possibilities for recording heritage values and allowing research and analysis to be conducted on this data.

10.8 Managing Risk

Preserving space heritage for future generations need not increase the risks for functioning spacecraft, crewed habitats or future orbital space industries. All defunct satellites do not present equal collision or fragmentation risks. An analysis of the ECOB space debris index indicated that fragmentation tended to occur in the first few years after launch. For spacecraft in orbit for 20 years or more, fragmentation events rarely occurred (Letizia et al. 2017). This suggests that the older a satellite is, the less susceptible it is to fragmentation and the lower its contribution to the debris population. Areas of peak density, such the band 760–870 km, present greater collision risks. Retaining historic satellites in situ in such locations may be more difficult to justify. Any cultural heritage management plan for a significant spacecraft will have to incorporate satellite debris modelling and predictions of its orbit from decades to 100s of years into the future.

Heritage management is a process of balancing competing interests, and in this case, of balancing the intrinsic values of the spacecraft with the instrumental values of the space environment. It makes sense to draw on tested terrestrial strategies to achieve this balance. The case of Envisat shows some of the management options in the case where environmental harm outweighs cultural significance.

10.8.1 Envisat

There are a number of ways for calculating which defunct spacecraft present the greatest risks, but the ESA satellite Envisat always tops debris removal 'hit-lists'. Envisat was launched in 2002 for Earth observation, and is one of the largest spacecraft in orbit after the space stations. Contact with the satellite was lost in 2012, meaning it cannot be manoeuvred and a controlled re-entry is not possible; but it is likely to remain in orbit for 150 years. Two catalogued debris objects per year are estimated to pass within about 200 m of the satellite in its present orbit. According to some predictions, collisions with Envisat could trigger the self-sustaining cascade of collisions known as Kessler Syndrome (Kessler and Cour-Palais 1978). It does not seem likely that any level of cultural significance would outweigh these risks (but this cannot be assumed).

A cultural heritage management plan for the satellite would start with a significance assessment following the Burra Charter guidelines. The satellite in its current condition in orbit cannot be fully recorded, although it might be possible to use ground-based instruments to observe the condition of its external surfaces. What mechanism will be used to de-orbit it is unclear, but this is its likely fate. If the de-orbit is controlled by an external mechanism, it is likely to be targeted at the ocean 'satellite graveyard' located at Point Nemo. If this is not possible, the most durable components such as titanium or steel pressure vessels and fuel tanks might possibly survive on the surface. Like Skylab they may become collectible souvenirs of space, but could also be collected for display in a museum.

Mitigation strategies are limited in the case of Envisat. In terrestrial contexts, the parameters of a road or power line may be changed to avoid damage to heritage. Managing Envisat is more in line with remediation of contaminated land. No significant fabric can be kept in its present location. However, there are options. A terrestrial parallel, more often used to preserve environmental values, is a heritage off-set. The loss of values in one area is off-set by retaining heritage with similar or equivalent significant fabric or related significance in another category. Unlike the numerous telecommunications satellites of similar design, Envisat is a one-off. There is nothing else like it.

Off-sets are less widely applied in heritage management as, unlike ecologies or vegetation, heritage is non-renewable. Without the physical satellite itself, its traces on Earth become more significant. Following the Dublin Principles and in accordance with the significance assessment, an off-set could consist of efforts to gather and curate associated documentation, components, models or prototypes relating to the satellite; and to record oral histories. Such a strategy ensures that information about the satellite remains for those in the present and future who seek to understand twenty-first century human activities in orbit.

10.9 Conclusions

In terrestrial environmental management, heritage is just one component of mitigating the adverse impacts of development on the environment. Typically, environmental impact studies include flora, fauna, biodiversity, water, soil, air quality, noise, employment, health and community. Few of these apply to space as it is currently theorised. Because space is not seen as a system of interconnected parts, in which an impact on one can have impacts on others – ie a non-biological ecology – I argue that concepts of environment in space are too 'thin'. There have been studies considering the impacts of rocket exhaust on the upper atmosphere (eg Voigt et al. 2013), and of space junk on the ability of humans to use orbital space (eg Gupta and Roy 2018). But the SSA model is no longer sufficient to encompass what orbital space has become. There may be nothing living that we know of in Earth orbit – apart from the crew of the International Space Station – but it is a complex environment encompassing space weather, cosmic rays, solar wind and flares, magnetic fields, animal bodies and human burials, and the human imagination from the Brick Moon to Starlink. One way to characterise this is as a cultural landscape (Gorman 2005a). Humans living in space, of whom it is assumed that ISS crew are only the beginning, will not be dealing with a situation. They will be living in an environment. The orbital environment has its own aesthetic qualities and affordances for machines and robots. There are smells (gunpowder) and perspectives (the Earth from outside) that are unique to space and contribute to the human experience of this environment.

The Anthropocene, a geological era shaped and defined by human activities upon Earth, has not yet been officially recognised; however, it has proven exceptionally powerful in theorising the world humans now live in. From the flatness of the cultural landscape, we can extend our conceptions of the geometric volume around Earth into an Anthropocene manifestation (Gorman 2014). As spacecraft fragment and decay, terrestrial elements and minerals are becoming part of the background soup of near-Earth space. Heritage management of space junk makes most sense as a component of a broader framework which does not separate 'natural' from 'cultural' and which acknowledges the synergistic effects of all cultural, chemical, mechanical and electromagnetic contributions. Indeed, at the level of the dust that is being created by the decay of spacecraft, it may be difficult to distinguish natural particles from those created by human activity without capture and microscopic examination. Both will become components of the space environment (Gorman 2019b).

Orbital space has never been conceived of as a pristine wilderness with value in its own right. Indeed, our knowledge of its state prior to human intervention is imperfect; the early scientific satellites sent data back to us about the space environment even as they were in the process of changing it. But we will never, now, return Earth orbit to a pre-Space Age 'emptiness'. It is becoming increasingly necessary to expand concepts of management and sustainability in outer space beyond the 'situation'.

References

American Institute of Aeronautics and Astronautics. (n.d.). AIAA Historic Aerospace Sites – Igor I. Sikorsky Memorial Airport, Connecticut. https://vtol.org/files/dmfile/AIAA-Sikorsky-Airport-VS-3001.pdf

Anglin, R. (2008). The World Heritage List: Bridging the cultural property nationalism-internationalism divide. *Yale Journal of Law & the Humanities, 20*, 241–274.

Anselmo, L., & Pardini, C. (2016). Ranking upper stages in Low Earth Orbit for active removal. *Acta Astronautica, 122*, 19–27.

Aplin, G. (2002). *Heritage, identification, conservation and management*. Melbourne: Oxford University Press.

Augé, M. (1992). *Non-places: An introduction to an anthropology of supermodernity*. Le Seuil: Verso.

Australia ICOMOS. (2013). *The Burra Charter*. http://australia.icomos.org/publications/charters/

Barclay, R. L. and Brooks, R. (2002). In situ preservation of historic spacecraft. *Journal of the British Interplanetary Society, 55*(5/6), 71–181.

Capelotti, P. J. (2010). *The human archaeology of space. Lunar, planetary and interstellar relics of exploration*. Jefferson: McFarland and Company.

Chairman of the Joint Chiefs of Staff. (2018). *Space operations Joint Report 3–14*. https://www.jcs.mil/Portals/36/Documents/Doctrine/pubs/jp3_14.pdf.

Commonwealth of Australia. (2017). *List of overseas places of historic significance to Australia*. http://www.environment.gov.au/heritage/places/list-overseas-places-historic-significance-australia

European Space Agency. (2020, February). *Space debris by the numbers*. https://www.esa.int/Safety_Security/Space_Debris/Space_debris_by_the_numbers

European Space Agency. (n.d.) *DISCOS*. https://discosweb.esoc.esa.int/web/guest/home

Fewer, G. (2002). Towards an LSMR and MSMR (Lunar and Martian Sites and Monuments Records): Recording the planetary spacecraft landing sites as archaeological monuments of the future. In M. Russell (Ed.), *Digging holes in popular culture: Archaeology and science fiction* (pp. 112–120). Oxford: Oxbow Books.

Fredericks, K. (1993, September 15). Community radio from space. *Green Left Weekly*. https://www.greenleft.org.au/content/community-radio-space

Gorman, A. C. (2005a). The cultural landscape of interplanetary space. *Journal of Social Archaeology, 5*(1), 85–107.

Gorman, A. C. (2005b). The archaeology of orbital space. In *Australian space science conference 2005* (pp. 338–357). Melbourne: RMIT University.

Gorman, A. C. (2009a). The cultural landscape of space. In A. Darrin & B. L. O'Leary (Eds.), *The handbook of space engineering, archaeology and heritage* (pp. 331–342). Boca Raton: CRC Taylor and Francis Press.

Gorman, A. C. (2009b). The gravity of archaeology. *Archaeologies, 5*(2), 344–359.

Gorman, A. C. (2009c). The archaeology of space exploration. In D. Bell & M. Parker (Eds.), *Space travel and culture: From Apollo to space tourism* (pp. 129–142). Malden: Wiley- Blackwell/The Sociological Review.

Gorman, A. C. (2014). The Anthropocene in the solar system. *Journal of Contemporary Archaeology, 1*(1), 87–91.

Gorman, A. C. (2015). Robot avatars: The material culture of human activity in Earth orbit. In B. L. O'Leary & P. J. Capelotti (Eds.), *Archaeology and heritage of the human movement into space* (pp. 29–47). Heidelberg: Springer.

Gorman, A. C. (2016a). Culture on the moon: Bodies in time and space. *Archaeologies, 12*(1), 110–128.

Gorman, A. C. (2016b, July 30). The death of a spacecraft. *Day of Archaeology*. https://www.dayofarchaeology.com/the-death-of-a-spacecraft/

Gorman, A. C. (2017). Not all space debris is junk: A comprehensive management strategy for culturally significant spacecraft. *68th International Astronautical Congress (IAC), Adelaide, Australia, 25–29 September 2017. IAC-17, A6,8,2,x40612.*

Gorman, A. C. (2019a). *Dr Space Junk vs the universe: Archaeology and the future*. Cambridge MA: MIT Press.

Gorman, A. C. (2019b, September 12). Cosmic sweep. *The Wheeler Centre*. https://www.wheelercentre.com/notes/cosmic-sweep

Gorman, A. C., Wallis, L., & Burke, H. (Eds.). (2018). *Road tripping: A journey of artefacts*. Adelaide: Wallis Heritage Consulting.

Gupta, B., & Roy, R. (2018). Sustainability of outer space: Facing the challenge of space debris. *Environmental Policy and Law, 48*(1), 3–7.

Harrison, R., & Schofield, J. (2010). *After modernity: Archaeological approaches to the contemporary past*. Oxford: Oxford University Press.

Heritage Council of NSW. (2001). *New South Wales historical themes*. https://www.environment.nsw.gov.au/resources/heritagebranch/heritage/themes2006.pdf

Högberg, A., Holtorf, C., May, S., & Wollentz, G. (2017). No future in archaeological heritage management? *World Archaeology, 49*(5), 639–647.

Holtorf, C. (2001). Is the past a non-renewable resource? In R. Layton, P. Stone, & J. Thomas (Eds.), *Destruction and conservation of cultural property* (pp. 286–295). Walnut Creek: Left Coast Press.

ICOMOS-TICCIH. (2011). *Joint ICOMOS-TICCIH principles for the conservation of industrial heritage sites, structures, areas, and landscapes "The Dublin Principles"*. https://ticcih.org/about/about-ticcih/dublin-principles/

Justus, J., Colyvan, M., Regan, H., & Maguire, L. (2009). Buying into conservation: Intrinsic versus instrumental value. *Trends in Ecology & Evolution, 24*(4), 187–191.

Kessler, D. J., & Cour-Palais, B. G. (1978). Collision frequency of artificial satellites: The creation of a debris belt. *Journal of Geophysical Research, 83*(A6), 2637–2646.

Klinkrad, H. (2015). Hazard of orbital debris. In J. N. Pelton & F. Allahdadi (Eds.), *Handbook of cosmic hazards and planetary defense* (pp. 808–834). Cham: Springer.

Labadi, S. (2007). Representations of the nation and cultural diversity in discourses on World Heritage. *Journal of Social Archaeology, 7*(2), 147–170.

Letizia, F., Colombo, C., Lewis, H., & Krag, H. (2017). *Extending the ECOB space debris index with fragmentation risk estimation*. Paper presented at the 7th European Conference on Space Debris.

Lipe, W. (1984). Value and meaning in cultural resources. In H. Cleere (Ed.), *Approaches to the archaeological heritage: A comparative study of world cultural resource management systems* (pp. 1–11). Cambridge: Cambridge University Press.

Mace, O. (2017). Australis Oscar 5: *The story of how Melbourne University Students built Australia's first satellite*. Adelaide: ATF Press.

McDowell, J. C. (2020). The Low Earth Orbit satellite population and the impacts of SpaceX Starlink constellation. *The Astrophysical Journal Letters, 892*(2), I.36.

NASA. (2020). *The Artemis Accords – Principles for a safe, peaceful, and prosperous future*. https://www.nasa.gov/specials/artemis-accords/index.html

NASA Orbital Debris Program Office. (2020). *Orbital debris frequently asked questions*. https://orbitaldebris.jsc.nasa.gov/faq/#

O'Leary, B. L. (2006). The cultural heritage of space, the Moon and other celestial bodies. *Antiquity*, Project Gallery. http://www.antiquity.ac.uk/projgall/oleary/

Philo, S. (2014). *British invasion: The crosscurrents of musical influence*. New York: Rowman and Littlefield.

Power, A., & Smyth, K. (2016). Heritage, health and place: The legacies of local community-based heritage conservation on social wellbeing. *Health and Place, 39*, 160–167.

The Hague International Space Resources Governance Working Group. (2019). *Building blocks for the development of an international framework on space resource activi-*

ties. https://www.universiteitleiden.nl/binaries/content/assets/rechtsgeleerdheid/instituut-voor-publiekrecht/lucht%2D%2Den-ruimterecht/space-resources/bb-thissrwg%2D%2Dcover.pdf

The Outer Space Institute. (2020). *The Vancouver recommendations on space mining*. http://www.outerspaceinstitute.ca/docs/Vancouver_Recommendations_on_Space_Mining.pdf

Voigt, C., Schumann, U., Graf, K., & Gottshcaldt, K. (2013). Impact of rocket exhaust plumes on atmospheric composition and climate — An overview. *Progress in Propulsion Physics, 4*, 657–670.

Westacott, E. (2019). Intrinsic vs. instrumental value – A basic distinction in moral philosophy. *ThoughtCo*.https://www.thoughtco.com/intrinsic-and-instrumental-value-2670651

Westwood, L., Gibson, G., O'Leary, B., & Versluis, J. (2010, January 30). *Nomination of the Objects associated with Tranquility Base to the California State Historical Resources Commission*. Accepted by unanimous vote to the California State Register of Historical Resources on 30 January 2010.

Zurlini, G., Petrosillo, I., & Cataldi, M. (2008). Sociological systems. In *Encyclopedia of ecology* (pp. 3264–3269). New York: Academic.

Chapter 11
Changing Space Security Dynamics and Governance Debates

Rajeswari Pillai Rajagopalan

Abstract Space has become an inalienable aspect of our daily lives. Yet, there is little debate on ways to keep it safe, secure and sustainable. A number of space security developments including the development of counter-space technologies, proliferation of small satellites and the US-China space competition have impeded an effective outer space regime. But the biggest challenge continues to be the lack of consensus among major space powers that have hindered the process of developing effective space security measures. This paper calls for a fresh approach to space security governance by seeking a pro-active role for the middle space powers and to look for threat-specific solutions rather than overarching agreements which are difficult to develop under the current international political climate.

11.1 Introduction

Technological advancements in outer space have permeated everyday lives today, be it the use of GPS (Global Positioning System) for navigational purposes, meeting the growing communication requirements or weather forecasting. As opposed to the Cold War days when space was primarily used by the military for strategic purposes, today militaries around the world have developed greater reliance on space for a number of conventional military applications. Despite this large-scale use of space across different sectors, there is inadequate appreciation for the steps required to keep space safe, secure and sustainable. This is not because there is lack of understanding among the space powers in how space can be protected and safeguarded. Rather, major power politics has impeded the rule making exercise including reviewing the existing mechanisms. There are also newer technologies such as for satellite inspection, refueling and repair (or broadly the on-orbit satellite servicing activities) meant to clean up space, but which arouse suspicion given the

R. P. Rajagopalan (✉)
Observer Research Foundation, New Delhi, India
e-mail: rpr@orfonline.org

M. de Zwart, S. Henderson (eds.), *Commercial and Military Uses of Outer Space*,
Issues in Space, https://doi.org/10.1007/978-981-15-8924-9_11

current state of international politics and the state of rivalry among big powers. Rules of engagement for these sets of applications are yet to be developed but it seems unlikely that we will see much progress in the near term.

These trends play out most seriously in the Indo-Pacific region. The region is already going through major geopolitical shifts and uncertainties due to power transition in the region and beyond. While the US continues to maintain a nearly unbeatable edge in its military and technological power, the absolute power of the US has seen a gradual decline with the rise of other powers. This dynamic has been most evident in the case of US-China competition across different sectors, including in outer space. China's rise as a peer competitor to the US has created its own impact in the Indo-Pacific, giving way to more competition, rivalry and a budding arms race. The terrestrial politics and rivalry are evident in outer space as well (Pekkanen 2015). An additional worry is the early trends we see of a shift from space militarization to space weaponization. This is dangerous because it is a slippery slope that can provide imperative for a large number of players to weaponize outer space without considering the larger implications of their own actions. With the competition already increasing among the top space players such as China, Russia and the US, it is a matter of time before outer space goes from a peaceful to militarized environment to possibly even a domain of active warfare.

While geopolitics has contributed to the increasing security challenges in outer space, the domain has also seen a spurt in the number of actors and correspondingly, an increase in space activities as well. A domain that was dominated by two big players during the Cold War decades – the US and the USSR – has more than 80 players today. Along with the increase in the number of actors, there has also been a diversification of actors. Space activities are no more undertaken by just state players but private, commercial sector actors and even educational institutions have become active in many countries. In May 2020, SpaceX launched American astronauts into space in its Dragon Endeavor spacecraft, a demonstration of the extent of private sector involvement in space activities that were traditionally undertaken solely by state agencies. More actors and activities increase the potential for collision and other accidents in space. The crowding and congestion in outer space also increases the potential for radio frequency interference in space.

On the other hand, the growing geopolitical rivalry also has increased the likelihood of intentional attacks carried out by states as a way of denying advantages accrued through space during conflicts. These could manifest in cyber and electronic warfare in space, making safe, secure, and sustainable uses of outer space challenging. Of these many challenges, I would argue that the most serious one is developing effective global governance mechanisms. It is easy to find technical solutions to many of the problems in outer space, but the real difficulty lies in the current international political climate which has impeded the prospects of developing consensus among major space players. This has been particularly so in the last few years, with the latest casualty being the UN Group of Governmental Experts (GGE) on Prevention of Arms Race in Outer Space (PAROS).

The state of the regime with regard to outer space is a direct result of the changing geopolitical dynamics that is underway both in the Indo-Pacific and beyond. Most of the rising powers today are in Asia. But Asia itself is a divided house because of multiple, divergent interpretations of history, territory, and sovereignty. Prospects for stability in Asia are beginning to look grim in many ways including through acquisition of hard power capabilities by many countries. This, of course, extends to outer space too.

This chapter will outline key factors that are increasing the concerns regarding space security and how global governance in the area of space security needs to be strengthened to ensure safe, secure and sustainable use of outer space. These factors include crowded and congested space, growth of counter-space capabilities, US-China competition and its impact on global governance and lastly the growth of small satellite constellations. Given the big geopolitical shifts underway in the Indo-Pacific, this chapter will approach the space security issues primarily from an Asian perspective. The global governance section looks at the inadequacies of the existing mechanisms and suggests possible steps for addressing the lack of progress that is currently dominating these debates.

11.2 Key Factors to Be Considered in Space Governance and Security

There are a number of important contextualizing factors that are affecting both the space security dynamics and the global governance debates. While states have continued to reiterate that space must be used for peaceful purposes alone, in reality, all the major space powers are developing various capabilities for military uses of space. These are primarily driven by a form of security dilemma, the fear that inaction could cost their security. Every state that is capable of doing so therefore makes an effort to stay ahead and not let potential adversaries gain an advantage. Of course, as in all security dilemmas, this has fueled greater worry and wariness and suspicion among states, leading to a challenging climate when it comes to global governance. The next section will identify some of these factors that have made space security challenges particularly worse.

11.2.1 Crowded and Contested Space

While human activities in space go back several decades, many developing countries across different regions have only now begun to more fully understand how use of space can transform the lives of their people. This has led to a huge growth in space activities. Going forward, this will mean that there will be more countries joining outer space activities to get their share of the pie. This could lead to further crowding of outer space. But these countries, being latecomers, are naturally suspicious of any effort to curtail their access to outer space. This affects global governance efforts because even a perception of denial of space technology or rights to the newcomers will create fierce opposition. These states consider that space has been already polluted by the big players. They are unlikely to accept any restrictions now, when they enter space, in the name of global governance.

The number of countries in Asia with independent launch capabilities include China, India, Iran, Israel, Japan and North Korea (Lele 2012). There are, in addition, a number of emerging space players in the region including Australia, Bangladesh, Indonesia, Iran, Israel, Malaysia, North Korea, Pakistan, Singapore, South Korea, the United Arab Emirates, and Vietnam (Moltz 2011). This demonstrates both the extent of interest among the Asian countries and the growing sophistication of the region in terms of their space prowess. Not only states, but private players are also becoming active in the Indo-Pacific region. Traditionally, private sector participation was limited to mostly western countries. But lately, there has been a shift in Asia with many countries seeing the emergence of private space sector. China took the lead in this, and India and Japan have in limited numbers. China's LinkSpace, LandSpace and i-Space have become quite prominent in the Chinese space programme with impressive support from the Chinese government (Beall 2019; Rajagopalan 2019a, b, c, d).

The growing number of private players in space, of course, would mean the overall number of objects in space would go up, including active space objects, old and worn-out spacecraft and upper stages of launch vehicles and other debris. One immediate and direct fallout of a crowded space environment is the consequence for space debris and space traffic management, both of which will impact on the sustainable use of outer space. This is a problem that needs to be addressed in an effective manner because space, being a dual-use domain, has enormous benefits in the social and developmental sectors that are becoming increasingly critical for a number of countries. The growing military and security dimensions of space activities are a threat to the well-being of these countries and their population, even if they are not direct participants in such conflicts in space.

Space is today crowded and congested not only because of the rapid increase in the number of satellites but also that of space debris, which has increased the dangers of operating in space because of the higher risks of collisions in orbit. According to NASA, as of 1 January 2020, there are more than 23,000 orbital debris larger than 10 cm, and around 500,000 debris pieces that are in the size range between 1 and 10 cm in diameter. There are more than 100 million pieces of debris that exceed 1 mm size. The total amount of material orbiting the Earth as of 1 January 2020 exceeds 8000 metric tons (Orbital Debris Program Office. Frequently Asked Questions). Debris pieces are estimated to be travelling at about 17,500 miles per hour, which means even a small piece of debris can cause enormous damage to a satellite or spacecraft. This poses a particular problem for the International Space Station (ISS). So far, there have been very few incidents that have threatened the ISS or other satellites and resulted in collision, except for a French satellite that was damaged by debris from a disintegrated Ariane rocket stage in 1996 (Reichhardt 2008). A more high-profile incident was in 2009 when the US Iridium satellite was hit by a defunct Russian satellite, causing at least 2000 pieces of trackable debris pieces (NASA 2013). The gravity of the challenge can be seen from the number of total operational satellites that stands at 2666 as on 31 March 2020, a majority of which belongs to the US at 1327 followed by China at 363 and Russia at 169 (Union of Concerned Scientists 2020).

While the number of new players in space cannot be restricted, the big players in space need to find a way to restrict the use of space for a variety of military and defence applications, which is fuelling the competitive aspects of outer space activities. Given the complex geopolitics of the Indo-Pacific region, space has also become a contested realm and it is difficult to assume that states will stop using space for military and security requirements.

There are also problems related to radio frequency interference which are the result of the crowded and congested nature of outer space. On the other hand, such interference incidents could also be a result of intentional attacks as well efforts to jam or block radio signals (Powell et al. 2018; Weeden 2013). States have to acknowledge the fact that space is a limited commodity and unless safeguarded, uninterrupted use of space will be impeded. Therefore, corrective steps need to be undertaken especially with regard to space security activities because there are both demand and supply dynamics that affect the space security changes. Growing economic well-being and technological advancements in general allow for states to spend more resources on space. The demand side of the equation is somewhat more complicated and there are several competitive forces that drive space programmes. But if one country feels the need to develop certain space or counter-space capabilities as a way of developing deterrence, there will be pressure on other states to follow suit. The net result is negative for all and space is likely to be more polluted and access to space can be hampered.

11.2.2 Growth in Counter-Space Capabilities

Counter-space capabilities seek to deny adversaries any advantage from using space capabilities. This is a growing reality that a lot of spacefaring countries are beginning to confront. The increasingly destructive nature of engagements in the form of cyber and electronic warfare in space in addition to the development and testing of anti-satellite (ASAT) weapons contribute to rising anxieties and insecurities in outer space. While counter-space technologies are not entirely new, today's geopolitical scenario is such that states are demonstrating greater willingness to develop, demonstrate and even employ some of these capabilities against adversaries.

There are both kinetic and non-kinetic capabilities in this regard. Kinetic capabilities are one where there is physical destruction of a space object, but research and development of kinetic capabilities are not entirely difficult to hide from the larger community. Similarly, cyber and electronic warfare capabilities are harder to detect at the stage of development but are even more problematic, as attribution is difficult even when they are used to attack space assets. It is even difficult to ascertain, for instance, if a satellite is malfunctioning due to a mechanical failure or because of an intentional directed action by an adversary. These capabilities have been developed, tested and deployed in a number of instances. Growing instances of blinding or dazzling of satellites using laser, engaging in cyber warfare resulting in temporary disruptions, denial of services, and/or interfering in command and control systems and logistics networks are particularly concerning because these early trends are already pushing more countries to invest and develop such means as a way of securing themselves (The Economist 2019). Cyber and electronic warfare options have been found to be a lot more attractive because they are cheaper and therefore affordable, the biggest attraction being that it gives one the option of plausible deniability. These trends in developing counter-space capabilities such as cyber and electronic means to create disruptions are already confirmed to be in operation with some Indo-Pacific powers (Rajagopalan 2019a, b, c, d; Weeden and Samson 2019; Harrison et al. 2019).

Some of the instances demonstrating the use of cyber and electronic technologies in space include:

List of Incidents Involving Cyber and Electronic Warfare[1]

Country	System/Capability/Incident
US	Counter Communications System, an operational electronic warfare system developed by the US that is deployable across the globe to undertake uplink jamming against geostationary communication satellites and was operationalized in 2004.
US	Reported to have developed the technical capacity to jam Global Navigation Satellite System (GNSS) receivers such as GLONASS or Beidou, in a small restricted area of operation to limit the possibility of use by adversaries.
China	Media reports indicate that "several Chinese scientists claimed to have successfully blinded a satellite in a 2005 test using a '50–100 [kilowatt] capacity mounted laser gun in Xinjiang province'." (Fisher Jr 2017).
China	In an incident in 2006, China has reportedly attempted to blind US spy satellites that were flying over Chinese territory using high-powered lasers but the effectiveness of it is not known (Muradian 2006). While the authenticity of these reports may be an issue, US officials have claimed that China has this capability and has "exercised it"(Harris 2006).

(continued)

[1]The list of incidents is compiled from a number of different sources, including in an earlier paper by the author

Country	System/Capability/Incident
China	In 2011, a US–China Economic and Security Review Commission report stated that two US satellites were compromised in 2007 and 2008 through a ground station in Norway. The attack was traced to China although the US Government did not come out openly accusing it but said that the nature of the attack was linked to Chinese hackers. The attack was alarming because, in the 2008 attack in particular, the hackers managed to complete all the required steps to command the satellite, though no harm was done. Potentially, the hackers could have stolen data, redirected the solar panel array in ways that would result in damage, or even moved the satellite (Wolf 2011).
China	In 2014, the US National Oceanic and Atmospheric Administration confirmed that one of its satellites had been hacked. Though none of its data was compromised, published news reports blamed China (Capaccio and Bliss 2011).
Iran	In 2009, Iran had reportedly engaged in electronic warfare activities. During the Iranian presidential elections in 2009 and the Arab Springs demonstrations in 2011, Iran is believed to have jammed some of the news broadcasts such as the BBC's Persian TV so as to restrict the western media to the Iranian people (Halliday 2012). Reports continue to cover Iran's interference with commercial communications satellites thus preventing foreign broadcasts in Iran.
Iran	In 2011, Tehran is reported to have shot down a US RQ-170 UAV, "by jamming its satellite communications links and spoofing the GPS signals it received," however, these claims were not confirmed by the US government (Peterson and Faramarzi 2011)[a].
North Korea	Between 2010 and 2012, the Democratic People's Republic of Korea was accused of jamming the Republic of Korea's GPS signals for days at a time, affecting many planes, ships and personal devices (Williams 2012; BBC News 2012; Waterman 2012).
Russia	According to reports, Russia has completed the development of a laser-based ASAT on the A-60 aircraft, named 1LK222 Sokol Eshelon. These systems are meant to both dazzle and blind sensors on satellites. Equipped with adequate power, they can damage light- or heat-sensitive physical components on satellites (Harrison et al. 2018; Mathew 2018; Cenciotti 2016).
Russia	Reports indicate that Russia also developed two jammers, the R-330Zh and R-381 T2. These along with four other jamming systems, were used to jam GPS signals in Ukraine in 2014 (Harrison et al. 2018).

[a]In another specific incident reported by Christian Science Monitor in 2011, according to an unnamed European intelligence source, Iran had "managed to 'blind' a US satellite by 'aiming a laser burst quite accurately'.

As for ASAT capabilities, several tests have been conducted by China, the US, Russia and India. India became the fourth country after the US, Russia and China to have a demonstrated capability but there are possibly many more contenders who have either developed the capability or are in the process of developing. Given the growing sense of competition, rise of military space programmes and the early trends towards space weaponisation, one will see the unfortunate reality of several more powers developing such capabilities in the future.

11.2.3 US-China Space Competition

Even though the US continues to be the world's most powerful country in absolute terms, there has been a relative decline in its power and influence and this has had a determining impact on the current state of the global norms and rules for outer space. The fact that global governance debates continue to be divided between two schools of thought, one led by the US and the other by the China-Russia combine, has not been helpful in developing the necessary compromise to make progress on several important space security issues. The lack of consensus among the major powers, driven by these two camps, has been demonstrated ample times, most recently at the 2018–19 UN Group of Governmental Experts (GGE) on PAROS (Prevention of Arms Race in Outer Space). The divisive politics had sharpened to such an extent that the 25 member States of the GGE could not even agree on a text and there was no consensus report from the GGE.

The widely prevalent disagreement among majors is a direct impact of the US-China rivalry that plays across many domains. The US continues to maintain a lead in the area of high technology including in space but China's rise has been impressive as well as consequential. China's steady mastery of disruptive technologies in space has been watched with concern and the fact that there will be implications for regional and global security cannot be ignored (Peoples 2013). China has made significant advances in the space realm including landing an unmanned mission on the near side of the Moon using Chang'e 3 in December 2013 and landing on the far side of the Moon using Chang'e 4 in January 2019 (Jones 2019a, b). China is possibly aiming for a human mission to the Moon next, with Beijing already making announcements about the possible establishment of a research base on the Moon. China is also on its way to establishing its own version of the International Space Station (ISS) in low earth orbit by 2024. This will be the time the current ISS will be winding up unless additional resources are available to keep it running (Howell 2020).

While these achievements are broadly in the civil space arena, China's activities have not been restricted to these alone. Its first successful anti-satellite (ASAT) weapon testing in January 2007 and the development of an array of counter-space capabilities are worrying both from an Indo-Pacific and the space sustainability perspective. With most militaries moving from a platform-centric to network-centric military operations, the reliance on space can prove to be a huge vulnerability to adversaries who are looking for ways to cripple the effectiveness of military operations. This is precisely what China has come to conclude looking at the US' heavy reliance on space-based network systems. So, on the one hand, China has been developing many of the space capabilities with the US in mind, but its actions have had many neighbours worried and forced them to respond so that they are not left lagging considering the national security complications. This has been so particularly because China has grown shown greater willingness to use these technologies in a disruptive fashion, triggering new competitive developments in the region and beyond.

In one of the first instances, China broke a decades-long unwritten moratorium on ASAT testing while simultaneously developing a range of other counter-space capabilities, making space significantly competitive and contested. The Chinese ASAT test spurred other countries to engage in such actions – the US' ASAT test called Operation Burnt Frost in 2008, and Indian ASAT test called Mission Shakti (Shakti means power in Hindi) in March 2019. The Indian ASAT test is a direct fallout of the Chinese ASAT test because India had shown little interest in ASATs prior to China's test (Rajagopalan 2011). Since China's ASAT test, India had to try to match up to a capability that China had already demonstrated both to develop its own deterrent capability and to protect its assets. Japan too is now moving in this direction. It has allocated funds for developing an interceptor to protect its own assets and strengthen its deterrence (The Japan News 2019). Japan has remained concerned about China's growing space prowess, including a satellite fitted with a robotic arm, whose purpose is still not very clear. This has compelled Japan to develop its own defense capabilities and institutional arrangements so as to protect itself in space. The Japanese government is yet to make a final decision on this but plans to do so by the end of the next (2021) financial year in order to launch the system by the mid-2020s. Countries like India and Japan that have had a purely civilian orientation to their space programmes, are having to respond to significant changes in the Indo-Pacific neighbourhood, a classic illustration of the security dilemma in operation (Johnson-Freese 2018).

Australia, another country affected by China's growing capabilities, has made no formal call on ASATs but the policy community is beginning to debate about the need for Canberra to develop it as a possible deterrent against the many vulnerabilities that are present in a heavily networked world today. Malcolm Davis of the Australian Strategic Policy Institute (ASPI) for instance uses the emerging space arms race as a context to suggest that "the cost of using space for benign and malign purposes is dropping, and there's an accelerating proliferation of space 2.0 technologies" (Kuper 2019). France is also taking some steps in that direction. France recently developed a defence space strategy that calls for the establishment of a "large Space Command" under the French Air Force (French Ministry of the Armed Forces 2019). French President Emmanuel Macron announced in July 2019 that the Command will come into existence on 1 September in Toulouse, south-west France. The Armed Forces Minister Florence Parly explained the decision from a self-defense angle stating that "If our satellites are threatened, we will blind those of our adversaries. We reserve the right to choose the time and means of the riposte: it could imply using powerful lasers deployed from our satellites or from patroller nano-satellites" (Mackenzie 2019). Therefore, countries both in the Indo-Pacific region and outside will witness significant space competition and these are mostly steered by terrestrial politics than ever before (Rajagopalan 2019a, b, c, d).

11.2.4 Proliferation of Small Satellites

A relatively new trend in space activities is the proliferation of small satellites – mini, micro and nano-satellites, by states as well as non-state actors, adding to the already significant challenge in the crowded and congested outer space realm. This is in line with the global trend where bigger satellite constellations are being broken into smaller ones. Advances in areas such as miniaturisation of electronics as well as mechanical parts, computing technologies and sensors and other key technologies along with the lower costs for satellites and satellite launch vehicles have aided this shift towards smaller satellites in the space market (Sami 2020). Small satellites are also favoured for another reason. Strategic factors have also been important, especially redundancy. The possibility of one's satellites being knocked out or damaged intentionally, especially during conflict, is a reality and small satellites are much easier to launch on demand and replace. The preparation time as well as the cost work out much better as compared to that of big satellites. Nevertheless, there are problems associated with small satellites, the biggest one being tracking and detection. But given the increasing utility of satellites in conventional military operations, small satellites are here to stay, and they make the congestion in outer space a bigger challenge in the coming years.

11.3 The Need for Strong Governance Measures

The various factors detailed in the previous section point to the urgent need to review existing governance mechanisms or establish new rules of the road so that space remains safe, secure and sustainable. But finding a right approach to address the challenges in space has not been an easy task. As mentioned earlier, the lack of consensus among major space players has been a major impediment in developing an effective outer space regime. In this section, I will briefly outline existing rules and agreements as well as their drawback, some current efforts to develop new rules and norms and suggest some alternative efforts.

11.3.1 Existing Measures and Gaps

Outer space is not completely bereft of any rules and regulations. There is a body of legal instruments that have regulated outer space activities for a few decades now. The Outer Space Treaty (OST) or more formally "the Treaty on Principles Governing the Activities of States in the Exploration and Use of Outer Space, including the Moon and Other Celestial Bodies" of 1967 is the foundation agreement that has been guiding outer space activities until now. It has been an effective measure in large measure in keeping the sanctity of outer space intact, but it is now showing its age. For instance, because the treaty came about in the security context of the Cold War, with only two major space powers, it prohibits the placement of weapons of mass destruction (WMD) in outer space. This addressed the threat apparent then but one which is no longer the same. There is no mention of conventional weapons, which represents the threat today. Given that outer space plays such an important role in conventional military operations, either the OST has to be reviewed or a new mechanism needs to be developed. There are other similar gaps and ambiguities in the OST (and in other treaties governing outer space) which, if not addressed, can be used by states to pursue their specific national security interests but to the detriment of the larger interest. These problems affect any new effort at developing treaties for outer space. For instance, just to give an example, there is ambiguity on the definitions of many key concepts: how does one define "peaceful use of outer space" or what is considered "weaponisation of space"? The lack of clarity and the lack of a commonly agreed definition, understanding and interpretation is an important obstacle for developing effective rules of the road.

The other legal instruments along with the OST are the Rescue Agreement of 1968, the Liability Convention of 1972, the Registration Convention of 1976 and the Moon Agreement of 1984. All the treaties except for the Moon Agreement have large numbers of signatories. Nevertheless, there are issues with compliance to these agreements. For instance, a fuller compliance of states towards their obligations to treaties by providing complete information (as they are required to) about their space activities can go a long way in removing the concerns and suspicions about their space activities. States are reluctant to embrace such transparency because it has obvious security implication, but the lack of transparency increases insecurity of others.

There have been efforts to keep afloat measures such as the Prevention of Arms Race in Outer Space (PAROS) for tackling the specific issue of arms race in outer space but the Conference on Disarmament (CD) in Geneva is yet to have a meaningful debate on PAROS. This also demonstrates the failure of some of the UN institutions pursuing security and arms control agenda. For example, the CD has not been able to conduct any substantive negotiations for over two decades because they cannot even agree on what issues to consider for negotiations.

11.3.2 Recent Efforts

States are mindful of the deteriorating space security scenario and there have been several proposals addressing various aspects of space security but none of them have been fruitful as yet. Some of the recent proposals include the Russia-China sponsored draft Treaty on the Prevention of the Placement of Weapons in Outer Space, the Threat or Use of Force against Outer Space Objects (PPWT), originally proposed in 2008 (with a revised text introduced in 2014); the EU-initiated International Code of Conduct for Outer Space Activities (2010), the 2013 UN Group of Governmental Experts on Transparency and Confidence Building

Measures (TCBMs), the 2018–19 GGE on further practical measures for the PAROS. The divisive politics among the major powers has sharpened almost to the point of no-return. As mentioned before, the latest GGE on PAROS is the best example demonstrating the uncompromising nature of major power politics. This is also because of the broader philosophical disagreements among states on the kind of governance measures to be pursued. The US-led western countries believe in non-legal political measures such as Transparency and Confidence Building Measures (TCBMs) as more feasible in the contemporary context (given the lack of trust between different countries) whereas a group of countries led by China and Russia articulate their preference for legally binding measures (Rajagopalan 2018).[2] Finding a middle ground between these two schools has been very challenging, and as yet, unsuccessful.

11.4 Time for Innovative Approaches

Given these seemingly intractable challenges, states have to become more innovative in addressing the growing space security threats. One suggestion might be to look at the possibility of developing threat-specific technical measures instead of looking at the most comprehensive measures. Reaching consensus on such overarching initiatives is unlikely to yield any positive results in the near term. Identifying threat-specific measures can possibly produce only limited success but it is an idea that is worth pursuing in the absence of support for the more comprehensive proposals.

In this context, it is also important to learn from past experience as we develop new measures. For instance, mechanisms akin to the Nuclear Non-Proliferation Treaty (NPT), which focused on technology denials to prevent the spread of nuclear weapons, need to be avoided both because technologies related to outer space have already spread. Moreover, given the extensive use of space in the civilian sectors, there will be little support for any such technology-denial regimes. If one is to look at the past mechanisms, it is useful to look at more practical instruments such as the Chemical Weapons Convention (CWC) or the Biological Weapons Convention (BWC) both of which dealt with technologies that had already spread, including to the private sector. Mechanisms need to be devised in a manner so as to monitor if states are diverting civil space capabilities for developing military space programs, though even this is not easy. Monitoring and verifying space capabilities can be a daunting challenge but future measures must be mindful of these requirements. Irrespective of the mechanism that takes shape, rules are to be framed for operations and activities than control and denial of technology.

[2]This source offers a perspective on how TCBMs can help generate greater confidence in each to work on legal measures.

A second approach worth examining is the role of middle powers in kickstarting debates on some of the tricky issues. Given the global as well as Asian power shifts that are underway, it is unlikely that major power politics will settle any time in the next couple of decades, which means that relying on major space powers to come up with possible solutions to space security challenges are unlikely to see any forward movement. But in the meantime, the security challenges are becoming more threatening. One could explore the role of middle powers in using evidence-based technical measures to ideate possible solutions. Such initiatives could be taken to a larger platform involving all the major space players at a later stage but it is important to survey the possibility of developing technology-aided solutions within a smaller grouping first. Some of the technically and financially capable middle space powers are Australia, Canada, India, Japan, South Korea and Switzerland. This group of countries can possibly fund technology-aided projects to address specific threats. One major advantage of such a grouping is that they need not be tied to the political and security compulsions of the major powers and can instead focus on space in more neutral ways. Monitoring and evidence-based verification work can be appealing to a number of countries, which could eventually build up support base among the great powers as well. Such efforts may not bring immediate results but they are important steps in building confidence among the great powers through evidence-based and technology-aided solutions. A related issue that needs to be discussed is the possible venue of future space security and arms control debates, in particular whether the CD is still the appropriate one.

There is little doubt that the situation is getting increasingly worrisome. Unless urgent action is taken, we may very well see outer space become not only an arena of competition in scientific endeavours but potentially a real battlefield as well.

References

BBC News. (2012, May 2). *North Korea Jamming Hits South Korea Flights*. http://www.bbc.com/news/world-asia-17922021

Beall, A. (2019, August 28). China's private space industry is rapidly gaining ground on SpaceX. *Wired*. https://www.wired.co.uk/article/china-private-space-industry.

Capaccio, T., & Bliss, J. (2011, October 27). Chinese Military suspected in Hacker Attacks on US Satellites. *Bloomberg*. https://www.bloomberg.com/news/articles/2011-10-27/chinese-military-suspected-in-hacker-attacks-on-u-s-satellites

Cenciotti, D. (2016, October 5). Russia has completed ground tests of its high-energy airborne combat laser system. *The Aviationist*. https://theaviationist.com/2016/10/05/russia-has-completed-ground-tests-of-its-highenergy-airborne-combat-laser-system/

Fisher Jr, R. D. (2017, February 23). China's progress with directed energy weapons. *US–China Economic and Security Review Commission* testimony. https://www.uscc.gov/sites/default/files/Fisher_Combined.pdf

French Ministry of the Armed Forces. (2019, July). Defence space strategy.

Halliday, J. (2012, March 14). BBC fears Iranian cyber-attack over its Persian TV service. *The Guardian*. http://www.theguardian.com/media/2012/mar/14/bbc-fears-iran-cyber-attack-persian

Harris, F. (2006, September 26). Beijing secretly fires lasers to disable US Satellites. *The Telegraph*. https://www.telegraph.co.uk/news/worldnews/1529864/Beijing-secretly-fires-lasers-to-disable-US-satellites.html
Harrison, T., Johnson, K., Roberts, TG. (2018, April). Space threat assessment 2018. *Center for International and Strategic Studies*. https://csis-prod.s3.amazonaws.com/s3fs-public/publication/180823_Harrison_Space ThreatAssessment_FULL_WEB.pdf
Harrison, T., Johnson, K., & Roberts, T. Space Threat Assessment 2019. *Center for security and international studies*. https://csis-prod.s3.amazonaws.com/s3fs-public/publication/190404_SpaceThreatAssessment_interior.pdf
Howell, E. (2020). International space station at 20: commercialization increases as end of life looms, August 28, 2020. https://www.space.com/international-space-station-20-years-commercialization.html/
Johnson-Freese, J. (2018, December 1). Asia's many space races. *The Diplomat*. https://thediplomat.com/2018/12/asias-many-space-races/
Jones, A. (2019a, January 3). Chang'e-4 returns first images from Lunar Farside following historic landing. *Space News*. https://spacenews.com/change-4-makes-historic-first-landing-on-the-far-side-of-the-moon/
Jones, A. (2019b, September 11). Chinese space station core module passes review but faces delays. *Space News*. https://spacenews.com/chinese-space-station-core-module-passes-review-but-faces-delays/
Kuper, S. (2019, May 24). Developing Australia's anti-satellite capabilities as a deterrent. *Defence-Connect*. https://www.defenceconnect.com.au/key-enablers/4109-developing-australia-s-anti-satellite-capabilities-as-a-deterrent
Lele, A. (2012, November 21). An Asian space race?. *Space News*. https://spacenews.com/an-asian-space-race/
Mackenzie, C. (2019, July 26). France plans to boost its self-defense posture in space. *Defense News*. https://www.defensenews.com/global/europe/2019/07/26/france-plans-to-boost-its-self-defense-posture-in-space/
Mathew, A. (2018, February 26). Russia completes development of Airborne Anti-satellite laser weapon. *DefPost*. https://defpost.com/russia-completes-development-airborne-anti-satellite-laserweapon/
Moltz, J. C. (2011). *Asia's space race: National motivations, regional rivalries, and international risks*. New York: Columbia University Press.
Muradian, V. (2006, September 25). China tried to blind US Sats with laser. *Defense News*. https://www.ar15.com/forums/general/Chi-na_Tried_To_Blind_U_S__Sats_With_Laser/5-501978/
National Aeronautics and Space Administration (NASA), Orbital debris program office. Frequently asked questions. https://orbitaldebris.jsc.nasa.gov/faq/
National Aeronautics and Space Administration (NASA). (2013, September 27). *Space debris and human spacecraft*. https://www.nasa.gov/mission_pages/station/news/orbital_debris.html#:~:text=On%20Feb.the%20inventory%20of%20space%20junk
Pekkanen, S. (2015, March 27). Asia's simmering rivalries are shifting to outer space, where anything goes. *Forbes*. https://www.forbes.com/sites/saadiampekkanen/2015/03/27/asias-simmering-rivalries-are-shifting-to-outer-space-where-anything-goes/#405f1feb2090
Peoples, C. (2013). Introduction: Reading East Asia's space security dilemmas. *Space Policy, 29*, 95–98.
Peterson, S., & Faramarzi, P. (2011, December 15). Exclusive: Iran Hijacked US drone, says Iranian engineer. *The Christian Science Monitor*. https://www.csmonitor.com/World/Middle-East/2011/1215/Exclusive-Iran-hijacked-USdrone-says-Iranian-engineer
Powell, T.D., Lubar, D.G., & Jones, K.L. (2018). Bracing for impact: terrestrial radio iInterference to satellite-based services. *Center for Space Policy and Strategy, The Aerospace Corporation*. January 2018. https://aerospace.org/sites/default/files/2018-05/BracingForImpact_0.pdf
Rajagopalan, R. (2011, November 10). India's changing policy on space militarization: The impact of China's ASAT test. *India Review, 10*(4), 354–378.

Rajagopalan, R. (2018, June 7). *TCBMs for a sustainable outer space*. Presentation to the Conference on Disarmament, Geneva. https://unog.ch/80256EDD006B8954/(httpAssets)/5D28719E45F62914C12582A500410385/file/180607-TCBMs-CD-Raji-Speaker+Notes.pdf
Rajagopalan, R. (2019a, May). *Electronic and cyber warfare in outer space*. Space Dossier 3 United Nations Institute for Disarmament Research. http://www.unidir.org/files/publications/pdfs/electronic-and-cyber-warfare-in-outer-space-en-784.pdf
Rajagopalan, R. (2019b, May). *Electronic and cyber warfare in outer space*. Space Dossier 3 United Nations Institute for Disarmament Research. http://www.unidir.org/files/publications/pdfs/electronic-and-cyber-warfare-in-outer-space-en-784.pdf
Rajagopalan, R., (2019c, July 19). The ISRO isn't enough. India needs its own Elon Musk or Jeff Bezos. *Quartz India*. https://qz.com/india/1668876/indias-quest-to-take-the-isro-to-the-top/
Rajagopalan, R. (2019d, August 23). China extends terrestrial rivalries into orbit with new space race. *Nikkei Asian review*. https://asia.nikkei.com/Opinion/China-extends-terrestrial-rivalries-into-orbit-with-new-space-race
Reichhardt, T. (2008, March). Satellite smashers. *Air & Space Magazine*. https://www.airspacemag.com/space/satellite-smashers-25098285/#:~:text=The%20first%2C%20or%20the%20first,at%20nine%20miles%20a%20second
Sami, A. (2020, May 25). *A shift from small constellation system of large satellites towards large constellation system of small satellites in space architecture*. Public University of Navarre (UPNA). https://traductordeciencia.es/a-shift-from-small-constellation-system-of-large-satellites-towards-large-constellation-system-of-small-satellites-in-space-architecture/
The Economist. (2019, July 18). *Attacking satellites is increasingly attractive—And dangerous*. https://www.economist.com/briefing/2019/07/18/attacking-satellites-is-increasingly-attractive-and-dangerous
The Japan News. (2019, August 19). *Satellite interceptor sought by mid-2020s*. https://the-japan-news.com/news/article/0005948349
Union of Concerned Scientists. (2020, April 1). *Satellite quick facts*. https://www.ucsusa.org/resources/satellite-database#.XAj4iC17HOQ
Waterman, S. (2012, August 23). North Korean Jamming of GPS shows system's weakness. *The Washington Times*. https://www.washingtontimes.com/news/2012/aug/23/north-korean-jamming-gps-shows-systems-weakness/
Weeden, B. (2013). Radio frequency spectrum, interference and satellites fact sheet. *Secure World Foundation*. https://swfound.org/media/108538/swf_rfi_fact_sheet_2013.pdf
Weeden, B., & Samson, V (Eds). (2019, April). Global Counterspace capabilities: An open source assessment. *Secure World Foundation*. https://swfound.org/media/206408/swf_global_counterspace_april2019_web.pdf
Williams, M. (2012, May 3). DPRK Jamming GPS signals, says Seoul. *North Korea Tech*. https://www.northkoreatech.org/2012/05/03/dprk-jamming-gps-signals-says-seoul/
Wolf, J. (2011, October 28). China Key Suspect in U.S. Satellite Hacks: Commission. *Reuters*. https://www.reuters.com/article/us-china-usa-satellite/china-key-suspect-in-u-s-satellite-hacks-commission-idUSTRE79R4O320111028

Chapter 12
"The Province of all Humankind" – A Feminist Analysis of Space Law

Cassandra Steer

Abstract This chapter argues that greater diversity is needed in the space sector, and this will only be achieved when women feel they are truly part of the structures and institutions that govern space. International space law today contains many powerful remnants of the Cold War era, including gender-specific language in the OST which states that space shall be "the province of all mankind". While we cannot change the terms of the treaty, we can change the way language is used today within the space governance power structures and in future legal instruments. Some may argue that this is not necessary, since "mankind" is intended to include women, however a feminist legal analysis reveals the power dynamics in processes of law-making, interpretation and application which maintain a status quo. In space law and governance, that status quo includes geopolitical dominance by a small group of elite countries, and underrepresentation of women at all levels of decision-making. In the space sector as a whole, the status quo is also a persistent lack of diversity, which not only limits design decisions, but can negatively impact our future in space. If we seek to build new human societies in space, on the Moon, and eventually on Mars, this requires reproductive health and the full participation of all members of these new societies. We cannot achieve this unless we radically shift the assumption of male biology and the male experience as the norm, and unless we move away from elite decision-makers forming laws and governance structures to their own benefit. This chapter undertakes to learn from achievements of feminist legal activism in domestic laws, and to apply this to the process of space law-making, interpretation and application. The use of language in the law is of paramount importance, and so a feminist outcome would be to actively re-interpret the OST and ensure that space becomes the "province of all humankind".

C. Steer (✉)
Australian National University, Canberra, Australia
e-mail: cassandra.steer@anu.edu.au

M. de Zwart, S. Henderson (eds.), *Commercial and Military Uses of Outer Space*, Issues in Space, https://doi.org/10.1007/978-981-15-8924-9_12

12.1 Introduction

Feminist legal scholarship has contributed to significant legislative reform throughout history, from the suffragette movement at the turn of the twentieth century, to laws covering labour and equal pay, equality in education, sexual misconduct, reproductive health, gender identity, and much more. Feminist scholarship in international law has also had an impact on the ways in which gender is framed in international criminal law, refugee and immigration laws, peacekeeping operations, international security and human rights law. Yet there are still power constructs in the ways in which international law is formed, interpreted and applied that reflect gender normativity, and which negatively impact the participation of women in these decision-making and law-making exercises. In international space law, we have inherited the legacy of the Cold War, a time when women were simply not at the treaty negotiating table, and our equality was not considered important in the international legal framework. The term used in the 1967 Outer Space Treaty (OST) that space shall be "the province of all mankind" is representative of this (Article I, OST 1967). This chapter will unpack the ways in which this gender determined language reflects uneven power constructs, and how these impact upon the participation of certain States as much as it does on women in the space sector and in space law-making.

This matters because diversity matters. Without the equal participation of women and of all nationalities in the development of the space sector, we cannot fulfill on the declaration of the OST that space shall be the province of all, and "shall be free for exploration and use by all States without discrimination of any kind, on a basis of equality and in accordance with international law." (Article I OST 1967)

Feminist legal theorists have demonstrated that law is neither neutral nor universally applicable, but rather is inherently political and an expression of power. Those who have power, determine the position of those who do not, and exclude those with less power from the law-making processes, thus cementing a status quo of inequality. The feminist method of discourse analysis can be used to discuss why language matters in the law. As lawyers, our words are our tools, and cases nearly always turn on the interpretation of a phrase, word or even a comma. The work of eminent feminist scholars is instructive as to the impact of gendered language in the law, and why it's important to reconstruct legal language in order to reconstruct social reality to be more equal and inclusive. From this, I will focus on the reasons why the term "the province of all mankind" is problematic in space law, and how we can reinforce greater equality and diversity in the space sector through employing and insisting upon gender neutral language, even if we cannot change the terms of the OST.

Space law has been described as "resistant" to feminist analysis or methodology (Harris Rimmer and Ogg 2019), and perhaps this is a result of the relatively small global community of space lawyers, among whom remains a predominance of an older generation, and of men, who generally understand space law to be neutral and universal, and do not see the "need" to apply a critical perspective. Meanwhile, critical legal scholars and practitioners who have impacted many areas of domestic and international law, remain relatively unaware of the pertinent issues of space law. The only noticeable critique of space law comes from Third World Approaches to International Law (TWAIL), which shares a similar agenda to feminist approaches: unmasking the political structures which lead to continued dominance by an elite group of states, and striving for distributively just solutions (Servant 2010; Eslava and Pahuja 2011; Gathii 2018, 2011; Aganaba-Jeanty 2016a, b, Steer 2019). In some ways, then, this chapter strives to answer the call for diffusion and mainstreaming of feminist critique into an area of international law where it is long overdue (Harris Rimmer and Ogg 2019).

12.2 Law and Power

Defining what a "feminist analysis" of space law entails is complex, because there is no single definition of "feminism", nor of feminist legal scholarship. There are many important differences between various feminist critiques of the law and legal processes, and feminist engagements with international law are sometimes quite apart from those with domestic law. For example, feminist legal theory at the national level began with litigation in many countries at the beginning of the twentieth century over whether women could be considered "persons" under various pieces of legislation determining who had the right to vote or be elected into political positions (Supreme Court of Canada 1928; Thornton 2019, p. 2). Throughout the twentieth century feminist critiques led to change in legislation to improve equality for women in the workplace and in education, protecting health and reproductive rights, and responding to gendered sexual harassment and violence (MacKinnon 1987, 1989; Dworkin 1993). Often these critiques have required a change in the legislative language used which referred more often than not to "he", or which assumed men's and women's experiences were the same, but there have been differences of opinion among feminist theorists as to whether formal equality between men and women should be sought, or special laws should be introduced to respond to women's different positions (Levit et al. 2016, p. 17). Critical race feminists later pointed out the need to include a wider range of voices and experiences, beyond the dominant white feminist perspective, (Hooks 1989, 1992; Williams 2000) since the beneficiaries of feminist legal reform tended to be white, middle-class women (Thornton 1986, p. 9).

Even with this wide range of "feminisms", at its core, feminist legal scholarship is concerned with the power constructs that shape the law, and the fact that we are not all equal before the law, despite this being one of the highest claims to moral and normative superiority that law makes. The ripple effect of feminist movements across society, and across different areas of domestic law in the twentieth century, inevitably impacted approaches to international law. Feminist method is particularly important when unveiling the power structures of international law. Critiques have emerged since the 1990s of the ways in which international law is formed, and who maintains positions of power in the negotiation, interpretation and application of international law.

The seminal article "Feminist Approaches to International Law" written by Hilary Charlesworth, Christine Chinkin and Shelley Wright, was the first thorough undertaking by international law scholars to critique the processes and impacts of international law through a feminist lens, and has remained the most cited article on the matter (Charlesworth et al. 1991). These authors pointed out that law claims a special authority based on the notion that it is an "autonomous entity, distinct from the society it regulates" and that it "operates on the basis of abstract rationality, and is thus universally applicable and capable of achieving neutrality and objectivity" (Charlesworth et al. 1991). However, law cannot be separated from the political, economic, historical and cultural context in which it operates, and in which the people who are subject to it live. Building on feminist scholarship of the decades preceding them, these authors highlight that just as women are impacted differently by domestic laws, so too the organisational structure of international law reflects a male perspective and ensures its continued dominance.

> States are patriarchal structures not only because they exclude women from elite positions and decision-making roles, but also because they are based on the concentration of power in, and control by, an elite and the domestic legitimation of a monopoly over the use of force to maintain that control. (Charlesworth et al. 1991)

Thus, the traditional doctrine of sources is itself a mechanism for asserting objectivity, neutrality and rationality, and therefore asserting the authority of law and of the State entities that form the law. Yet the law-making processes according to each of those sources are the result of a system that ensures the most powerful states remain the most influential. The people who are in the seats of power determine the content of that influence, through asserting structures and determining the negotiating outcomes for treaties and customary law. Treaties, though negotiated in a multilateral setting, are always the result of political give and take, and the States who can leverage their power the most, certainly take more than they give.

The formation of the core space law treaties is a clear expression of these power dynamics. Negotiated and signed between 1967 and 1979, the content of the core space treaties is dominated by the interests of the USSR and the U.S. at the time, since they were the two key players in the space race, and they both perceived the constraints that they agreed to as being more applicable to their opponent than to themselves (von der Dunk 2015). Both these countries provided the drafts for the treaties, and were at the forefront of all negotiations, based on their specific interests (Moltz 2014). It is for this reason that many describe the 1967 OST as predominantly an arms control treaty (Moltz 2014). However, developing nations from the African, Asian and Latin American regions, many of which were in the throes of asserting their independence through decolonisation, brought a different set of concerns to the negotiating table. They saw the need to be able to access and use these new technologies for their economic and social advancement, and feared that space benefits would be limited to the same small group of elite, developed nations who dominated world politics in all other areas (Jankowitsch 2015). They were therefore able to successfully influence some of the principles included in the treaty, to ensure that space shall be "for the benefit and in the interests of all countries, irrespective of their degree of economic or scientific development, and shall be the province of all mankind." However, these principles receive far less attention than the articles focused on arms control, or the question of liability, which were and remain the focus of the "greater" powers.

While feminist engagements with international law may not transform the system immediately, they can "unsettle its boundaries" (Harris Rimmer and Ogg 2019). Greater participation by women in seats of power, including those where the law is made and applied, is therefore necessary not only for including women's experiences in the law, but also as a step towards ensuring more inclusive perspectives on issues which may not, at first glance, appear to be issues of gender. This is an important lesson when turning to space law, the structures under which it is formed, and the states and people who hold the power to interpret and apply it.

12.3 Law and Power in Space

The unequal power dynamics of international law are made apparent through the critical lens provided by feminist and TWAIL methodologies. But even through slightly less critical lenses, many theories of international law-making will recognize that these processes respond to the demands of international relations at a given point in time, and to the shifts in power dynamics, as different issues or values come to the forefront (Boyle and Chinkin 2007; D'Aspremont 2010). The assertions of objectivity and neutrality made by the traditional doctrine of sources are therefore laid bare in all sub-branches of international law, including space law.

I have argued elsewhere that space law and space politics are determined by the same big players who dominate terrestrial geopolitics: in simple terms, the U.S. first, the EU second, with a counterweight from China and Russia (Steer 2019). As with any other area of international law and international relations, these powerful actors determine how issues are framed and what legal structures are (or are not) put in place. For example, in the year that the OST came into effect, Austria, Iran and Egypt proposed to the UN the establishment of a global space organisation, that could mimic the International Atomic Energy Agency, to monitor space activity and promote the spread of space technology to less-developed countries, in line with the notion that space should be for the benefit and in the interests of all (Moltz 2014). But because this was not in the interests of either of the super-powers, who had a monopoly over space and its governance and who were consumed with their race to the Moon, it did not gain any support from them.

Another example is the functioning of the UN Committee on Peaceful Uses of Outer Space (COPUOS), the body under which all five core space treaties were successfully and rapidly negotiated during the Cold War. For many years now,

COPUOS has been deadlocked on seeking multilateral solutions to current issues in space, such as the rising concern for weaponization, or access to and sale of natural resources in space. Part of the problem is that it is bound to make decisions by consensus – a process which was intended to ensure equality of all votes among nations, regardless of their relative power. But this only made sense at a time when there were fewer members, and a narrower range of concerns about activities in space, in the mid-twentieth century at the beginning of the space race (Lyall and Larsen 2018). Over time, with more members, it has become more difficult to come to consensus on any issues, and even the adoption of the non-binding Guidelines for the Long-Term Sustainability of Outer Space Activities took over a decade to agree upon (Working Group on the Long-term Sustainability of Outer Space Activities 2019; Martinez 2018). Often commentators will point to the fact that there have been no new space treaties since 1979, as evidence that there is no international appetite for new treaties (Aganaba-Jeanty 2016a; Meyer 2018). However as I have previously argued, this represents only the view of certain influential players, namely the U.S. and the allies which are tethered to it for national security or economic reasons: including Australia, Canada, much of the EU and the U.K. (Steer 2019). When the U.S. takes a position, tethered allies will follow suit.

Issues where the U.S. position has thoroughly dominated include the rejection of a proposed Treaty for the Prevention of the Weaponization of Outer Space (PPWT), and resolutions before the UN General Assembly on the Prevention of an Arms Race in Outer Space (PAROS). China and Russia have co-sponsored draft PPWT texts before the Conference on Disarmament, but the U.S. has consistently rejected them, due largely to the fact that these proposals come from its greatest opponents. In 2018, the U.S. was the only country to vote no to all four resolutions proposed at the UN General Assembly on PAROS (Report of the First Committee 2018; Meyer 2018), and in 2019, the U.S. opposed discussions in a dedicated UN Group of Government Experts on PAROS, where there was general agreement about the need for an arms control treaty for space (Patriota 2019; Steer 2019).

The U.S. position also clashed with the position of the EU on the tail end of the negotiating process working towards a non-binding International Code of Conduct for Outer Space Activities (ICoC) in 2015 (*Proposal for an International Space Code of Conduct, Draft* 2014). Despite having supported the initiative at its inception in 2008, the U.S. was unable to force the inclusion of a provision on the right to use force in self-defence, and as a result decided to jettison its support (Johnson 2014; Meyer 2015). The ICoC also faced pushback from the Global South, but for different reasons. The ICoC had originated as an EU initiative, and in its earliest iterations many developing countries objected to the Euro-centric procedures and selection of issues. To its credit, the EU responded to these critiques with a series of open-ended consultations in various regions of the world, before inviting all members of the UN to attend an international negotiation process in 2015. However, the concerns about process had not been addressed to the satisfaction of Brazil, Russia, India, China and South Africa (BRICS), who stated jointly that "the elaboration of such an instrument should be held in the format of inclusive and consensus-based multilateral negotiations within the framework of the UN" (*BRICS Joint Statement* 2015). These countries took the opportunity to assert some power over a process that typically represented the power imbalance of international law and international relations, and as a result the EU process – and ultimately the ICoC – failed.

These examples demonstrate the importance of taking into account the current realities of international relations and international law when seeking to govern space in the twenty-first century. As feminist critiques have shown, law cannot be separated from the political, cultural, economic, and historical context in which it plays out, and in which the nations and people exist who are affected by the law. Already in 1968, the Secretary of the UN at the time, U Thant, warned "that the space age is increasing the gap between the developed and developing areas of the world at an alarming rate" (Note by the Secretary General 1968) It is clear from the processes outlined here that certain countries are unwilling to let the power imbalance remain the status quo. This should be applauded, because that status quo keeps many nations out of the space race altogether, and limits the benefits they receive from human activities in space, in spite of the promise of the OST that space shall be the province of all.

It is clear, therefore, that the pretenses of international law as being neutral, objective, and universal are false, and that space law is as much an expression of power dynamics as is any other area of law. There is no equality between countries, despite the notion of formal equality as a value underpinning international law, and the status quo is determined by interests of a small handful of countries which have managed to institutionalize the power they held at the close of the Second World War. There is no equal access to space, nor is there distribution of the benefits derived from space, despite this being a promise of the OST. Space is far from being the "province of all mankind".

Indeed, space is even further from being the province of all *humankind*. Access to, benefits from, and governance over space is the province of an elite few, and within those few there is a gender imbalance which mirrors the geographical imbalance. At the time that the OST was drafted, not only were there no women at the negotiating table, but under the U.S. programme, women were excluded from being able to become astronauts. To become an astronaut, one had to be a military test pilot, a profession from which women were banned (Koren 2017). There was a strong lobbying campaign, led by highly qualified women pilots, to convince NASA and the White House to allow women to become astronauts (Klein 2017), and a clandestine "Women in Space" programme was bankrolled by the pioneering pilot Jackie Cochran (Weitekamp 2004). In this program, a number of women were selected by Dr. Randolph Lovelace, a contractor to NASA who led the physical tests and training for astronauts, to undergo the exact same training as the men, because he suspected women would be better candidates for space travel, due to our generally lighter weights and lower need for oxygen. A higher percentage of women passed the tests than men, and many of the women performed better than the male trainee astronauts. However, despite the test results, the deeply engrained sexism of the time prevailed. Apparently Lovelace's motives may have been focused on the need for women as secretaries and assistants in future long-term space habitations (Weitekamp 2004). When "Women in Space" candidate Jerrie Cobb testified before a congressional subcommittee in 1962, she stated "we seek, only, a place in our nation's space future without discrimination" (Klein 2017), but astronaut John Glenn testified that creating a programme to train women astronauts would compromise the race to land on the Moon before the Soviets. Moreover, he argued "the men go off and fight the wars and fly the airplanes and come back and help design and build and test them. The fact that women are not in this field is a fact of our social order." (Weitekamp 2004; Klein 2017). Ultimately the lobbying campaign failed, and the Women in Space program was shut down because NASA did not sponsor it.

One year later, the first woman in space was a Soviet woman, Valentina Tereshkova, in March 1963. The Soviets had beaten the Americans in yet another milestone in the space race, ostensibly breaking the glass ceiling for women's participation. However, she was not to be followed by another woman until 1982, when Svetlana Savitskaya flew on a mission to the Soviet Salyut Space Station. Upon her arrival, Savitskaya was handed an apron by her crewmates, who "joked" that she should get to work in the kitchen. Despite this rude welcome, she went on to perform a series of highly skilled engineering tasks for which she had been trained, including testing a tool for welding in space, and becoming the first woman to undertake a spacewalk (Lewis 2018).

Women are still vastly underrepresented in all STEM careers, and in the entire space sector generally, as well as at all international negotiating tables and in national law-making. It matters, then, a great deal, who has the power to determine the laws and norms applicable to human activity in space. If we are at all serious about the promises of the OST, then this power balance must shift. We must take into account the interests of many more players than just the most geopolitically influential as we seek new space law and governance solutions to today's and tomorrow's space activities. It starts with making explicit that space is not at all "the province of all mankind", let alone the province of all humanity.

12.4 Why Language Matters

By stating that there is a difference between space being a province "of all mankind" versus the province "of all humankind" I am asserting that language truly matters. Obviously, we cannot change the terms of the treaty, however in order to fulfill the object and purpose of the OST, we must do more than simply assume "all mankind" is an inclusive phrase, when everything discussed here shows that it is not. If the language is exclusive, then so is the reality. To rectify the inherent power imbalance in space law, space governance and so many aspects of the space sector, we must be proactive in addressing this language in all the other places it appears. The repetition in popular media of the term "province of all mankind" goes hand in hand with the use of terms such as "manned flight", "unmanned spacecraft", "manmade objects". How does this speak to girls wanting to take part in the space sector, or in the law-making and policy-making of the future? As Sally Ride, the first American woman astronaut, says: "you can't be what you can't see" ("Sally Ride" 2012). This language is much more exclusive than some are willing to accept. Dr. Alice Gorman, a renowned space archeologist, has put it perfectly:

> When you're a bloke, terms such as "mankind" automatically include you. You don't have to think about it at all; you're already in there. Now we all know that these terms are *supposed* to also include women; but the reality is a bit different. Firstly, women have to "think themselves into" such expressions, even if it happens at a subconscious level. Secondly, there have been studies which show that men tend to assume such expressions to refer to them alone and do not automatically include women unless stated, again often at a subconscious level. And finally, there are plenty of examples of women attempting to exercise a right of "man," only to be told it *does not apply* to them. So basically, the continued use of manned, mankind, etc, simply reinforces the impression that space is for men and not women. (Gorman 2014)

The notion that women have to "think themselves into" masculine expressions captures perfectly the motivations for litigating any number of cases over the decades that have challenged the use of gendered terms in the law. It is not enough to say that we can just assume "mankind" includes women, because in reality it does not. Even Neil Armstrong's famous words when he first stepped foot on the Moon applied automatically to men and mankind, while women were literally excluded from taking the small step of entering his profession until decades later, and no woman has yet been able to take the giant leap of reaching the Moon. It is also not enough to argue that Armstrong's *intention* was to be inclusive, or that the drafters of the OST did not *intend* to exclude women from this universal province. This is not about ill intention; it is about the creation and maintenance of a status quo that is built upon power dynamics and exclusion. Often, subconscious bias can be even more insidious and powerful because of its impact.

The exclusionary impact of the term "mankind" was felt even in the revolutionary moment of the *Déclaration des droits de l'homme et du citoyen* issued in 1789 by the French National Assembly – in English "Declaration of the rights of man and of the citizen" – to reflect the ideals of the French revolution and establish them as part of a new societal order. (English translation: *Avalon Project – Declaration of the Rights of Man – 1789*, n.d.; *Déclaration Des Droits de l'Homme et Du Citoyen de 1789*, n.d.) While the French language uses the masculine as the collective norm by default, the impact of this otherwise revolutionary document and the institutions it championed was to explicitly exclude women from these political and civil rights. Feminist playwright Olympe de Gouges responded with the *Déclaration des droits de la femme et de la citoyenne* – the Declaration of the rights of woman and of the female citizen (*Déclaration des droits de la femme et de la citoyenne*, n.d.; Warman 2016) – in which she followed the text of the original very closely, but replaced all terms relating to men or the male gender with "woman" or the female gender. Thus, the determination that "[t]he representatives ["*représentants*" in the masculine form] of the French people, organized as a National Assembly . . . set forth in a solemn declaration the natural, unalienable, and sacred rights of man", became: "The mothers, daughters and sisters, who make up the female representatives [*représentantes*] of the Nation, demand to be constituted as a National Assembly... [and declare] that all women are born and remain free and equal in rights". The right of the woman citizen to "participate personally or through her representative" asserted the right to political participation, not otherwise granted by the ostensibly universal language of the original declaration.

Olympe de Gouges proved too subversive even for the revolutionaries, and she was sent to the guillotine in 1793, because she "forgot her natural place" (Bar 2017). As a playwright and activist, she was cognizant of the power of language as a tool to shape societal beliefs and structural realities. Despite de Gouge's eloquent work, up until the twentieth century, the concept of "citizen" was coterminous with "man" and "masculinity", and as Ngaire Naffine points out, was "associated with actual men who monopolised the institutions of public power" (Naffine 1995). Women were not seen as being fit for public life, and were relegated to the private sphere, a distinction which many feminist theories highlight (MacKinnon 1987; Thornton 1995, 1991; Knop 1993; Butler 2011). Public law and governance have been very much the province of "mankind".

This is where the feminist method of discourse theory is so important. Discourse analysis studies language in relation to its social context, its purposes and effects. It "unmasks and disrupts the sameness implied by the universal language" that is so often used to assert equality and inclusion in a system where there is inherent inequality and exclusion (Thornton 2000). Ostensibly universal terms in the law are never truly universal, and Margaret Thornton's seminal work critiquing the "benchmark man" in legal language goes to the heart of this. Men do not need to prove their equality, or argue that they are included in any group that derives its rights or status from the law, because it's always implicit. However women, African Americans and First Nations peoples around the world have only gained citizenship and formal equality by fighting for it in the courts (Thornton 2000). And even then, their citizenship has been limited historically by not granting full political participation, and continues to be limited in many subtle ways by the legal definitions which continue to assert neutrality and universality despite reality being far from this.

As feminist legal scholars have taught us, nothing about the formation, interpretation nor application of the law is neutral, and the use of language in the law is key to forming and maintaining a deliberate status quo. Language is the tool of the legislator, the judge and the advocate, since entire cases turn on a single word, comma, or reconstructed meaning of a phrase. As Margaret Davies has elucidated, the "subject" of the law is based on the male experience, because it has been defined by it – literally, by the men in the seats of legal power (Davies 1994).

This is in line with linguistic theories which suggest that we construct our reality through language, since language is both formative and normative. Children learn to describe and interact with their surroundings by learning labels and concepts, and as adults we continue to do this and in doing so, are continually constructing our realities in a way that is anything but neutral or objective. As TWAIL feminist scholar Shelley Feldman points out:

> Knowledge is socially produced and embodies specific interests and historical circumstances. In this view, knowledge is situated and limited by its positioning ... (and) reflects our social experience, our understanding of our interests and our values. Different strategies of description and understanding are products of values and objectives created in a specific transitional network of power relation (Feldman 1998).

Knowledge is socially constructed through power relations. By applying discourse analysis, the power relations which determine the status of an individual – or a class of individuals – under the law become even more apparent. Law is not neutral, because the historical, social and political contexts in which legal language is formed and then applied through interpretation, are not neutral.

Feminist scholar Robin Lakoff has noted that "linguistic discrimination" was part of how women were denied access to power (Lakoff 1973). Yet we still see examples of linguistic discrimination in the very sources of language, such as the Oxford Dictionary. Recently, a dispute arose on Twitter about the consistently sexist examples given for the usage definitions of words, such that "doctor" was always given the male pronoun, whereas "nurse" was always given the female; and phrases derogatory towards women were used throughout, such as "the rising shrill of women's voices" for the word "shrill; and "nagging wife" for the adjective "nagging" (Flood 2016).

It's important to remember that even dictionary definitions are "selected by editors because they are used by specific populations, often dominant or powerful ones, in speech and in print" (Oman-Reagan 2016). Linguist Amanda Laugesen provides a fascinating historical account of what she calls "linguistic activism" in dictionaries, publication usage guides and political debates in Australia, the U.K and the U.S., and how this has impacted the way we speak and write in public media, educational institutions and society as a whole. She recounts debates surrounding whether changing language can create social change, or whether only the reverse is true (Laugesen 2019). Whereas Lakoff argued that only social change could create language change, and not the other way around, Dale Spender argued compellingly that a monopoly over language is one way that men of elite classes have maintained their primacy, and that changing both language and society are important tasks, to be achieved by eliminating sexist language (Laugesen 2019). Spender's work was influential, and Laugesen details a series of handbooks such as the 1980 "Handbook of Non-Sexist Writing for Writers, Editors and Speakers" published in British and Australian editions, and the "Guidelines for Equal Treatment of the Sexes in McGraw-Hill Book Company Publications", produced by the education publisher in U.S. and Australian editions in 1974 and 1980 (Laugesen 2019).

These handbooks were instrumental in changing the language used in television, radio and print media as well as in educational texts. The generic use of "mankind" was addressed in these handbooks, and central to debates that ensued. The McGraw-Hill Handbook suggested replacing "mankind" with "humanity" or "human race"; "manmade" with "artificial", "constructed", or other alternatives; "manpower" with "human power" or "human energy"; and suggested explicit, alternate uses of the pronouns "he" and "she" (Laugesen 2019). The handbooks, and the debates they sparked, also impacted official government language use. In

Australia, the Australian Council of Trade Unions introduced guidelines that stated the "importance of the educative role that the provision of non-sexist worded awards can have on men and women workers and employers", arguing that eradicating gender specific or sexist terms would bring equal pay and equal opportunity and treatment for women closer (Laugesen 2019). In 1986, the official Commonwealth Style Guide was updated by the Style Council – a powerful body that helps determine the use of language in the public sphere – to replace "Chairman" with "Chair" and "Ombudsman" with "Ombud". In her address to the Council, Anne Summers, First Assistant Secretary to the Office of the Status of Women, stated that "[w]e believe encouraging people to use different language will encourage them to think differently about women —including the way women think of themselves" (Summers 1986; Laugesen 2019). Similarly, when Senator Andrew Murray called for an amendment to a Bill to change the term "Chairman" to "Chair", he argued that language is "a very potent force of both oppression and change. How we use language sends messages about what sort of society we are" (*Parliamentary Debates* 1997; Laugesen 2019). Policies on gender-neutral language entered law schools around the world in the 1990s and 2000s, teaching future legal practitioners the importance of language in their profession.

Some resisted these changes as unnecessary political correctness, and argued in parliamentary debates that language is "natural" and could not be artificially changed – and that therefore terms such as "mankind" and "Chairman" are natural, and must be seen to be inclusive because of their long history (Laugesen 2019). Others argued that even if it could be changed, this would lead to lack of clarity and unwieldy sentences in legislation, however the Office of the Parliamentary Council stated unambiguously in its 2016 Drafting Guide:

> Gender-neutral language is used to make the language of legislation more inclusive… We have found that this approach does not make provisions more cumbersome or lead to a proliferation of "his and hers" etc. throughout new legislation as drafters tend to use the description of the relevant person (e.g. "the taxpayer's") instead. In many cases this makes the provision clearer than if just "his" was used and may also avoid ambiguity where there is more than one person referred to in the provision (Office of the Parliamentary Council 2016).

NASA's own style guide was not updated until 2006 to recommend non-gender specific language, such as replacing "manned" and "unmanned" with "human, piloted, unpiloted, robotic" (NASA 2006). However, as anthropologist Oman-Reagan observes, the gendered terms still appear in headlines and in some U.S. government institutions (Oman-Reagan 2016). He argues that we need to "unman" the language of space to reflect today's reality and invite more inclusivity in the future – and he accurately points out that the only truly "unmanned" space flight to date was Valentina Tereshkova's flight in 1963.

When "unmanning" space law, we cannot change the wording of the OST. However, we can ensure that we do not reify the language of the 1960s, and the inequality that it underscores. When referencing the treaty, if we are not citing the exact language of Article I, we should consistently use inclusive language – in doing so we pro-actively re-interpret the document's effects, and construct a reality that is more inclusive. It may even be useful to turn to some of the other official translations of the OST, such as French in which Article I states that the exploration and use of space "*sont l'appanage de l'humanité tout entiére*" – the prerogative of the entirety of humanity – or Spanish, in which they "*incumben a toda la humanidad*" – shall touch all humanity. Since even in official translations there is an act of interpretation, which is as much strictly linguistic as it is cultural, the notion of there being universal meaning in exclusionary language comes undone entirely. Thornton puts it eloquently:

> So well entrenched is the notion of maleness as universal throughout all branches of knowledge that special efforts by feminists would seem to be necessary to overcome its distorting effects.... Feminism, therefore, seeks to create a new vision of the world from the perspective of women: it is not content to accept the male standard as an objective and universal truth which purports to encompass all.... Feminist scholarship, then, aims to be "perspective transforming" (Thornton 1986).

12.5 Conclusion: The Province of [Hu]Mankind

It is clear, then, that language matters in how we construct reality. It is also clear that language matters especially in law, because it is a tool of power that can be wielded to oppress individuals and groups and deny rights, or to ascribe status and guarantee rights. If we want to ensure that the access to and use of space truly is for the benefit of all humankind, the prerogative of all humankind, and touches all humankind, then we need to widen the lens linguistically, as a starting point to broadening the reality. If we can "unman" the language of space law, and of space activities, then we can open the doors to more women, more nationalities, a greater diversity of identity, and therefore a richer and more realistic set of solutions to the problems we face as we seek to use and explore space further.

The first American woman astronaut, Sally Ride, experienced all kinds of sexism and ignorance, because she was the first exception to the assumed norm of the male body in space. She has recounted that she was asked by NASA engineers whether 100 tampons would be enough for her 2 week mission in space, and that an entire team had spent a great deal of time designing a make-up kit for her and the other women astronauts in training, assuming this would be among the essentials that these highly qualified astronauts would need in space (Ride 2002). Upon her return from her spaceflight, Ride described her frustration with the amount of press attention on the fact that she was the first American woman in space, and how she dealt with bathroom facilities and makeup, rather than on her accomplishments as an astronaut:

> They didn't care about how well prepared I was to operate the [Space Shuttle] arm or deploy communication satellites...It's too bad this is so important. It's too bad that we're not further along that it's a normal occurrence for a woman to go up on a Space Shuttle flight. It'll be a wonderful day when this isn't news."(Ride 2002)

Sadly, even in 2014, when Yelena Serova was about to become the first woman cosmonaut in 17 years, the press was obsessed with asking this space engineer about how she would keep her hairdo in space, and how she would manage being apart from her daughter during her mission. Her response was to the point: "aren't you interested in the hairstyles of my [male] colleagues?... My flight is my job. I feel a huge responsibility towards the people who taught and trained us." (France-Presse 2014). In 2015, when six women cosmonauts were about to undertake an eight day experiment living in a mock Russian spacecraft, to test the physical and psychological effects of long-term spaceflight, they were asked repeated questions by the press about how they would cope without makeup and shampoo (Grush 2015). And in March 2019, NASA planned and widely publicised the world's first all-female spacewalk, which was cancelled at the last minute as there was only one spacesuit available for a medium sized torso (Cantor 2019). A male astronaut had to take the place of Anne McClain, because he fit the larger suit. The problem was rectified by October 2019, and astronauts Christina Koch and Anne McClain performed the first "unmanned" spacewalk. But it is stunning that even a planned historical moment to signal improved gender quality in space was delayed due to a failure to think beyond the male body as the "norm" or standard. NASA points out that Koch and McClain graduated in a class of trained astronauts that was 50% women (Dunbar 2019), which is an excellent sign, however it's clear that more women are needed in the teams of on the ground engineers and decision makers as well.

As has been pointed out in feminist scholarship on international law:

> Long-term domination of all bodies wielding political power nationally and internationally means that issues traditionally of concern to men become seen as general human concerns, while "women's concerns" are relegated to a special, limited category. (Charlesworth et al. 1991)

This leads us to consider the underlying organising principles in the design and creation of long-term human habitation in space, human health in space, engineering of space systems, and decision-making and governance structures. If the organising principle is based on the assumption of the male norm, as it has been throughout history, then adding women after the fact to these design and governance decisions, as "the exception", leads to delays, increased costs, and new sets of problems. If ensuring the availability of a sufficient number of spacesuits for women astronauts is a problem because of a male organising principle, and if women are still not subjects of the same range and extent of health and medical testing for space programmes as men, how will we deal with the legal regulation of reproductive health in space, pregnancy and childbirth, the effects of medications on women's bodies rather than men's, the ways in which long-term habitation is designed for community living with a range of physical, emotional and psychological needs? Designing and governing new human societies in space, on the Moon and eventually on Mars, requires *a priori* reproductive health and the full participation of all members of these new societies. We cannot achieve this unless we radically shift the assumption of male biology and the male experience as the norm, and unless we move away from elite decision-makers forming laws and governance structures to their own benefit. If only an elite selection of the world's population is included in these design and decision-making processes, then only an elite few will benefit.

On the other hand, if "humankind" is represented, rather than "mankind", if the organising principle is inclusive and diverse to begin with, the entire range of solutions to all of these problems will be different. The same will go for nearer-term projects such as engineering decisions for robotics and exploration; governance decisions over our use of the natural resources in space, including the commercially valuable Low Earth and Geostationary orbits, the militarily valuable cis-lunar orbits (between the Earth and the Moon), and resources on the Moon; the integration of environmental principles into our expanded use of space for terrestrial benefits; and ensuring intergenerational distributive justice in the sense promised by the OST.

Like all legal systems, international law plays an important part in constructing reality (Charlesworth et al. 1991) – and therefore, so does space law with respect to the reality of the space sector. Feminist approaches aim to restructure traditional discourse and methodology to accommodate alternative world view, and as such they are about transformation rather than simply about technical law reform. These approaches could lead to the "creation of international regimes that revise notions of state responsibility, challenge the centrality of the state, and the traditional sources of international law" (Charlesworth et al. 1991).

If feminism has a power to transform legal method, it is because it permits us "a new way of seeing both the reality of our present lives and a new way of imagining a better one" (Mossman 1986). Feminist enquiries are about making visible the impact of patriarchal structures, but the ultimate objective is much larger than this: despite differences in approach, all feminisms are forward looking, striving to shape the future such that it is more inclusive, diverse, and fair in a distributive sense (Otto 2012). If space is to be truly accessible to, and for the benefit of, all humanity, then we need to ensure we are describing that reality accurately.

Space was the "province of mankind" throughout the twentieth century from the beginning of the space age – and even then only for mankind from the greater power nations. We are now well into the twenty-first century, where corporations and NewSpace technologies have disrupted the traditional national competition of power in space, ostensibly "democratising" space. Yet in space law and governance, there is still a status quo of geopolitical dominance by a small group of elite countries, and underrepresentation of women at all levels of decision-making, in such a way that future activities in space are unlikely to fulfil on the intention of the distributive equality envisaged by the OST, unless a transformative shift is undertaken. If we seek just distribution and equal participation in this new space race, we must be mindful of the ways in which we construct the political, economic and social realities, and proactive about the ways in which the law is used to articulate them. We must ensure that space is "for all humankind", in the truest sense of the word.

References

Aerberhard-Hodges, J. (2019). Practitioner perspective: Women and international treaty making – The example of standard setting in the International Labour Organization. In S. Harris Rimmer & K. Ogg (Eds.), *Research handbook on feminist engagement with international law*. Edward Elgar Publishers. https://www.e-elgar.com/shop/gbp/research-handbook-on-feminist-engagement-with-international-law-9781785363917.html.

Aganaba-Jeanty, T. (2016a). Introducing the Cosmopolitan Approaches to International Law (CAIL) lens to analyze governance issues as they affect emerging and aspirant space actors. *Space Policy, 37*, 3–11.

Aganaba-Jeanty, T. (2016b). Space sustainability and the freedom of outer space. *Astropolitics, 14*(1), 1–19.

Agreement Governing the Activities of States on the Moon and Other Celestial Bodies, United Nations Treaty Series vol. 1363, No. 23002 (1979).

Avalon Project – Declaration of the Rights of Man – 1789. (n.d.). Retrieved June 26, 2020, from https://avalon.law.yale.edu/18th_century/rightsof.asp

Bar, R. (2017, November 4). *Olympe de Gouges, the radical French feminist who was murdered twice*. Haaretz.Com. https://www.haaretz.com/world-news/europe/.premium.MAGAZINE-the-radical-french-feminist-who-was-murdered-twice-1.5462530

Boyle, A., & Chinkin, C. (2007). *The making of international law*. Oxford University Press.

BRICS Joint Statement Regarding the Principles of Elaboration of International Instruments on Outer Space Activities. (2015). https://perma.cc/5RZB-ZK57

Butler, J. (2011). *Gender trouble: Feminism and the subversion of identity*. Routledge.

Cantor, M. (2019, March 26). Nasa cancels all-female spacewalk, citing lack of spacesuit in right size. *The Guardian*. https://www.theguardian.com/science/2019/mar/25/nasa-all-female-spacewalk-canceled-women-spacesuits

Charlesworth, H., Chinkin, C., & Wright, S. (1991). Feminist approaches to international law. *American Journal of International Law, 85*(4), 613–645.

Chimni, B. S. (2006). Third world approaches to international law: A manifesto. *International Community Law Review, 8*(1), 3.

Convention on the Elimination of All Forms of Discrimination Against Women. (1981). United Nations, Treaty Series, 1249, p. 13. https://www.un.org/womenwatch/daw/cedaw/text/econvention.htm

Crawford, J. (2012). *Brownlie's principles of public international law*. OUP Oxford.

D'Aspremont, J. (2010). International law-making by non-state actors: Changing the model or putting the phenomenon into perspective? In M. Noortmann & C. Ryngaet (Eds.), *Non-state actors dynamics in international law – From law-takers to lawmakers*. Ashgate. http://papers.ssrn.com/abstract=1621764.

Davies, M. (1994). *Asking the law question*.

De Alwis, A. (2016, October 6). *Interview with Navi Pillay: Former UN High Commissioner for Human Rights*. Diplomatic courier. https://www.diplomaticourier.com/posts/interview-navi-pillay-former-un-high-commissioner-human-rights

Déclaration des Droits de l'Homme et du Citoyen de 1789. (n.d.). Retrieved June 26, 2020, from https://www.legifrance.gouv.fr/Droit-francais/Constitution/Declaration-des-Droits-de-l-Homme-et-du-Citoyen-de-1789

Déclaration des droits de la femme et de la citoyenne. (n.d.). Les Essentiels Littérature. Retrieved June 26, 2020, from https://gallica.bnf.fr/essentiels/anthologie/declaration-droits-femme-citoyenne-0

Dunbar, B. (2019, October 17). *Friday's all-woman spacewalk: The basics* [Text]. NASA. http://www.nasa.gov/feature/fridays-all-woman-spacewalk-the-basics

Dworkin, A. (1993). Prostitution and male supremacy. *Michigan Journal of Gender & Law, 1*, 1.

Edwards v. A.G. of Canada, 86 UKPC (Privy Council 1930).

Eslava, L., & Pahuja, S. (2011). Between resistance and reform: TWAIL and the universality of international law. *Trade, Law and Development, 3*, 103.

Estrada-Tanck. (2019). A feminist human security-human rights Lens: Expanding Women's engagement with international law. In S. H. Rimmer & K. Ogg (Eds.), *Research handbook on feminist engagement with international law*. Edward Elgar Publishers. https://www.e-elgar.com/shop/gbp/research-handbook-on-feminist-engagement-with-international-law-9781785363917.html.

Executive Order on Encouraging International Support for the Recovery and Use of Space Resources. (2020). (testimony of Donald J. Trump). https://www.whitehouse.gov/presidential-actions/executive-order-encouraging-international-support-recovery-use-space-resources/

Facts and figures: Peace and security. (n.d.). UN Women. Retrieved July 15, 2020, from https://www.unwomen.org/en/what-we-do/peace-and-security/facts-and-figures

Feldman, S. (1998). *Conceptualizing change and equality in the 'third World'Contexts*. Women in the Third World: An Encyclopedia of Contemporary Issues, (pp. 24–36).

Fineman, M. A., Jackson, J. E., & Romero, A. P. (2016). *Feminist and queer legal theory: Intimate encounters, uncomfortable conversations*. Routledge.

Flood, A. (2016, January 25). Sexism row prompts Oxford Dictionaries to review language used in definitions. *The Guardian*. https://www.theguardian.com/books/2016/jan/25/oxford-dictionary-review-sexist-language-rabid-feminist-gender

France-Presse, A. (2014, September 25). Female cosmonaut bats back questions about hair and parenting. *The Guardian*. https://www.theguardian.com/world/2014/sep/25/female-cosmonaut-russia-yelena-serova-press-questions-hair-parenting

Freeman, M. A. (2009). *Reservations to CEDAW: An analysis for UNICEF* (Policy and Practice, p. 48). UNICEF.

Gathii, J. T. (2011). TWAIL: A brief history of its origins, its decentralized network, and a tentative bibliography. *Trade, Law and Development, 3*, 26.

Gathii, J. T. (2018). *The agenda of third world approaches to international law (TWAIL)* (SSRN Scholarly Paper ID 3304767). Social Science Research Network. https://papers.ssrn.com/abstract=3304767

Gorman, A. (2014, September 6). How to avoid sexist language in space – Dr Space Junk wields the red pen. *Space Age Archaeology*. http://zoharesque.blogspot.com/2014/09/how-to-avoid-sexist-language-in-space.html

Gray, C. (2002). From Unity to polarization: International law and the use of force against Iraq. *European Journal of International Law, 13*(1), 1–19.

Gray, C. (2018). *International law and the use of force*. Oxford University Press.

Grush, L. (2015, November 1). *Russian female astronauts asked how they would cope in space without men and makeup*. The Verge. https://www.theverge.com/2015/11/1/9648718/russian-space-agency-women-cosmonauts-moon-mission-sexism

Hamilton, S. (2021). Lessons from Canada's famous five for Today's world changers. In C. Leprince & C. Steer (Eds.), *Women, peace and security: Feminist perspectives on international security studies*. McGill Queens University Press.

Harris, A. P. (1990). Race and essentialism in feminist legal theory. *Stanford Law Review*, 581–616.

Harris, A. P. (1999). Gender, violence, race, and criminal justice. *The Stanford Law Review, 52*, 777.

Harris Rimmer, S., & Ogg, K. (Eds.). (2019). *Research handbook on feminist engagement with international law*. Edward Elgar Publishers. https://www.e-elgar.com/shop/gbp/research-handbook-on-feminist-engagement-with-international-law-9781785363917.html.

Hooks, B. (1989). *Talking back: Thinking feminist, thinking black*. South End Press.

Hooks, B. (1992). *Yearning: Race, gender, and cultural politics. improvement in the status of women in the United Nations System: Report of the Secretary General*. (2019). UN General Assembly. https://undocs.org/en/A/74/220

Iraq war illegal, says Annan. (2004, September 16). *BBC News*. http://news.bbc.co.uk/2/hi/middle_east/3661134.stm

Jankowitsch, P. (2015). The background and history of space law. In F. von der Dunk (Ed.), *Handbook of space law*. Edward Elgar Publishing.
Jean Paul Akayesu. (1998, September 2). ICTR-96-4-T Judgment (International Criminal Court). https://unictr.irmct.org/sites/unictr.org/files/case-documents/ictr-96-4/trial-judgements/en/980902.pdf
Johnson, C. (2014). *Draft International Code of Conduct for Space Activities – Fact Sheet* [Fact Sheet]. Secure World Foundation. https://swfound.org/media/166384/swf_draft_international_code_of_conduct_for_outer_space_activities_fact_sheet_february_2014.pdf
Klein, C. (2017, July 24). Right stuff, wrong gender: The woman astronauts grounded by NASA. *History*. https://www.history.com/news/right-stuff-wrong-gender-the-woman-astronauts-grounded-by-nasa
Knop, K. (1993). Re/statements: Feminism and state sovereignty in international law. *Transnat'l L. & Contemp Probs*, *3*, 293.
Koren, M. (2017, March 23). *Why women weren't allowed to be astronauts*. Government Executive. https://www.govexec.com/technology/2017/03/why-women-werent-allowed-be-astronauts/136405/
Krause, J., Krause, W., & Bränfors, P. (2018). Women's participation in peace negotiations and the durability of peace. *International Interactions, 44*(6), 985–1016.
Lakoff, R. (1973). Language and woman's place. *Language in Society, 2*(1), 45–79.
Laugesen, A. (2019). Changing 'man made language': Sexist language and feminist linguistic activism in Australia. In *Everyday Revolutions*. ANU Press.
Levit, N., Verchick, R. R., & Minow, M. (2016). *Feminist legal theory: A primer* (Vol. 74). NYU Press.
Lewis, E. (2018, March 7). Svetlana Savitskaya – Unsung Hero of Space. The National Space Centre. https://spacecentre.co.uk/blog-post/svetlana-savitskaya-unsung-hero-space/
Lyall, F., & Larsen, P. B. (2018). *Space law: a treatise* (Second edition). Routledge.
MacKinnon, C. A. (1987). *Feminism unmodified*. Harvard University Press.
MacKinnon, C. A. (1989). *Toward a feminist theory of the state*. Harvard University Press.
Martinez, P. (2018). Development of an international compendium of guidelines for the long-term sustainability of outer space activities. *Space Policy, 43*, 13–17. https://doi.org/10.1016/j.spacepol.2018.01.002.
Meyer, P. (2015, August 31). Star-crossed: An international code of conduct for outer space? *OpenCanada*. https://www.opencanada.org/features/star-crossed-an-international-code-of-conduct-for-outer-space/
Meyer, P. (2018, December 11). Washington sparks a space spat at the United Nations. *Bulletin of the Atomic Scientists*. https://thebulletin.org/2018/12/washington-sparks-a-space-spat-at-the-united-nations/
Moltz, J. (2014). *Crowded orbits: Conflict and cooperation in space*. Columbia University Press.
Mossman, M. J. (1986). Feminism and legal method: The difference it makes. *Austl. JL & Soc'y, 3*, 30.
Naffine, N. (1995). *Sexing the subject (of law)*. Oxford University Press.
NASA. (2006). *Style Guide for NASA History Authors and Editors*. https://history.nasa.gov/styleguide.html
NASA. (2020). *NASA: Artemis Accords*. https://www.nasa.gov/specials/artemis-accords/index.html
Office of the Parliamentary Council. (2016). *OPC Drafting Manual*. https://www.opc.gov.au/sites/default/files/drafting_manual_1.pdf?acsf_files_redirect
Ogg, K. (2019). The future of feminist engagement with refugee law: From the margins to the Centre and out of the "pink ghetto.". In S. H. Rimmer & K. Ogg (Eds.), *Research handbook on feminist engagement with international law*. Edward Elgar Publishers. https://www.e-elgar.com/shop/gbp/research-handbook-on-feminist-engagement-with-international-law-9781785363917.html.
Okafor, O. C. (2005). Newness, imperialism, and international legal reform in our time: A TWAIL perspective. *Osgoode Hall Law Journal, 43*, 171.

Oman-Reagan, M. P. (2016, March 7). *Unmanning space language*. SAPIENS: https://www.sapiens.org/column/wanderers/outer-space-and-gendered-language/.

Otto, D. (2012). Feminist Approaches to International Law. In *The Oxford Handbook of the Theory of International Law*.

Parliamentary Debates. (1997). Senate of the Commonwealth of Australia. https://www.aph.gov.au/Parliamentary_Business/Hansard/

Patriota, A. (2019). *Chair's Summary: Open-ended intersessional informal consultative meeting on the work of the Group of Governmental Experts on further practical measures for the prevention of an arms race in outer space*. United Nations Group of Government Experts on the Prevention of an Arms Race in Outer Space. https://www.un.org/disarmament/wp-content/uploads/2019/03/paros-gge-open-ended-informal-consultative-meeting-chair-summary-final.pdf

Procedure for the nomination and election of judges of the International Criminal Court. (Resolution Resolution ICC-ASP/3/Res.6). (2004). International Criminal Court.

Proposal for an International Space Code of Conduct, Draft. (2014). (testimony of European Union). https://eeas.europa.eu/headquarters/headquarters-homepage/14715/eu-proposal-international-space-code-conduct-draft_en

Reference re meaning of the word "Persons" in s. 24 of British North America Act, (Supreme Court of Canada 1928).

Report of the Committee on the Peaceful Uses of Outer Space (A_7285E). (1968). United Nations General Assembly. https://www.unoosa.org/pdf/gadocs/A_7285E.pdf

Report of the First Committee. (2018). *Prevention of an Arms Race in Outer Space* (A/73/508). United Nations General Assembly Fourth Committee. https://undocs.org/A/73/508

Ride, S. K. (2002). *Nasa Johnson Space Center Oral History Project Oral History Transcript* (R. Wright, Interviewer) [Interview]. https://historycollection.jsc.nasa.gov/JSCHistoryPortal/history/oral_histories/RideSK/RideSK_10-22-02.pdf

Ride, S. (2012, September 1). *Harvard Business Review*. https://hbr.org/2012/09/sally-ride

Servant, V. (2010). Empowering the third world in international law formation: Comparing the declaration of the first meeting of equatorial countries of 1976 and part V and part xi of the UNCLOS III of 1982. *Cambridge Student Law Review, 6*, 117.

Sivakumaran, S. (2017). The influence of teachings of publicists on the development of international law. *International and Comparative Law Review, 66*, 1.

Status of Ratification Interactive Dashboard. (2020, July 14). UN Office of the High Commissioner of Human Rights. https://indicators.ohchr.org/

Statute of the International Court of Justice, 33 UNTS 993. (1946). https://www.icj-cij.org/en/statute

Steer, C. (2015, March 18). Four women at the top of the International Criminal Court – an international first. *IntLawGrrls*. https://ilg2.org/2015/03/18/four-women-at-the-top-of-the-international-criminal-court-an-international-first/

Steer, C. (2019). Who has the power? A critical perspective on space governance and new entrants to the space sector. *Georgia Journal of International and Comparative Law, 48*, 751.

Summers, A. (1986). *Inclusive language: Address to style council '86*. Australian National Dictionary Centre Archives.

Text – H.R.3979 – 113th Congress. (2013–2014). Carl Levin and Howard P. "Buck" McKeon National Defense Authorization Act for Fiscal Year 2015, (2014). https://www.congress.gov/bill/113th-congress/house-bill/3979/text

The Veto. (2015). Security Council Report, 3, 9.

The Veto : UN Security Council Working Methods. (2020, July 3). Security Council Report. https://www.securitycouncilreport.org/un-security-council-working-methods/the-veto.php

Thornton, M. (1986). Feminist jurisprudence: Illusion or reality. *Australian Journal of Law and Society, 3*, 5.

Thornton, M. (1991). The public/private dichotomy: Gendered and discriminatory. *Journal of Law and Society, 18*(4), 448–463.

Thornton, M. (1995). The cartography of public and private. *Public and private: Feminist legal debates*, (pp. 2–16).

Thornton, M. (2000). Towards strategic universalism. *Australian Journal of Legal Philosophy, 25*, 315.

Thornton, M. (2019). Challenging the legal profession a century on: The case of Edith Haynes. *University of Western Australia Law Review, 44*, 1.

Treaty on the Principles Governing the Activities of States in the Exploration and Use of Outer Space, including the Moon and Other Celestial Bodies, 610 U.N.T.S. 205 (1967). https://www.unoosa.org/oosa/en/ourwork/spacelaw/treaties/outerspacetreaty.html

True, J., & Riveros-Morales, Y. (2019). Towards inclusive peace: Analysing gender-sensitive peace agreements 2000–2016. *International Political Science Review, 40*(1), 23–40.

Valdes, F. (1997). Beyond sexual orientation in queer legal theory: Majoritarianism, multidimensionality, and responsibility in social justice scholarship or legal scholars as cultural warriors. *The Denver University Law Review, 75*, 1409.

von der Dunk, F. (2015). International space law. In F. von der Dunk & F. Tronchetti (Eds.), *Handbook of space law*. Edward Elgar Publishing.

Wald, P. (2011). Women on international courts: Some lessons learned. *International Criminal Law Review, 11*(3), 401–408.

Warman, C. (Ed.). (2016). Olympe de Gouges (1748–1793), Declaration of the Rights of Woman and the Female Citizen, 1791. In *Tolerance* (1st ed., Vol. 3, pp. 49–51). Open Book Publishers; JSTOR. https://www.jstor.org/stable/j.ctt19b9jvh.24

Weitekamp, M. A. (2004). *Right stuff, wrong sex: America's first women in space program*. JHU Press.

Williams, P. J. (2000). Race and rights. *Theories of Race and Racism: A Reader*. (pp. 421–430).

Working Group on the Long-term Sustainability of Outer Space Activities. (2019). *Guidelines for the long-term sustainability of outer space activities* (A/AC.105/C.1/L.366). Committee on the Peaceful Uses of Outer Space. https://undocs.org/A/AC.105/C.1/L.366

Index

M. de Zwart, S. Henderson (eds.), *Commercial and Military Uses of Outer Space*, Issues in Space, https://doi.org/10.1007/978-981-15-8924-9

GPSR Compliance
The European Union's (EU) General Product Safety Regulation (GPSR) is a set of rules that requires consumer products to be safe and our obligations to ensure this.

If you have any concerns about our products, you can contact us on

ProductSafety@springernature.com

In case Publisher is established outside the EU, the EU authorized representative is:

Springer Nature Customer Service Center GmbH
Europaplatz 3
69115 Heidelberg, Germany

www.ingramcontent.com/pod-product-compliance
Ingram Content Group UK Ltd.
Pitfield, Milton Keynes, MK11 3LW, UK
UKHW020817290926
15470UKWH00003B/1214
* 9 7 8 9 8 1 1 5 8 9 2 3 2 *